Classic chemistry experiments

Compiled by Kevin Hutchings
Teacher Fellow
The Royal Society of Chemistry
1997–1998

ROYAL SOCIETY OF CHEMISTRY

Classic chemistry experiments

Compiled by Kevin Hutchings

Edited by Colin Osborne and John Johnston

Designed by Imogen Bertin and Sara Roberts

Published by The Royal Society of Chemistry

Printed by The Royal Society of Chemistry

Copyright © The Royal Society of Chemistry 2000

Apart from any fair dealing for the purposes of research or private study, or criticism or review, as permitted under the UK Copyright Designs and Patents Act, 1988, this publication may not be reproduced, stored, or transmitted, in any form or by any means, without the prior permission in writing of the publishers, or in the case of reprographic reproduction, only in accordance with the terms of the licences issued by the Copyright Licensing Agency in the UK, or in accordance with the terms of licences issued by the appropriate Reproduction Rights Organisation outside the UK. Enquiries concerning reproduction outside the terms stated here should be sent to The Royal Society of Chemistry at the London address printed on this page.

For further information on other educational activities undertaken by The Royal Society of Chemistry write to:

The Education Department
The Royal Society of Chemistry
Burlington house
Piccadilly
London W1V OBN

Information on other Royal Society of Chemistry activities can be found on its websites:
www.rsc.org
www.chemsoc.org

ISBN 0 85404 9193

British Library Cataloguing in Publication Data.

A catalogue for this book is available from the British Library.

Classic chemistry experiments

RS•C

Contents

Foreword iv

Introduction v

How to use this book vi

Adapting the worksheets – examples vii

The role of information and communications technology (ICT) xv

Using the publication on the web xv

List of experiments xvi

List of experiments by categories. xviii

Health and safety xxv

Acknowledgements xxvii

Bibliography xviii

Experiments 1

1. Separating a sand and salt mixture 2
2. Viscosity 4
3. Rate of evaporation 6
4. Chromatography of leaves 8
5. The energetics of freezing 11
6. Accumulator 13
7. Electricity from chemicals 15
8. Iron in breakfast cereal 18
9. Unsaturation in fats and oils 22
10. The pH scale 23
11. Preparation and properties of oxygen 25
12. Identifying polymers 27
13. Energy values of food 32
14. A compound from two elements 35
15. Chemistry and electricity 37
16. Combustion 39
17. Determining relative atomic mass 41
18. Reaction of a Group 7 element (iodine with zinc) 44
19. Reactions of halogens 46
20. Sublimation of air freshener 49
21. Testing the pH of oxides 52
22. Exothermic or endothermic? 54
23. Water expands when it freezes 57
24. Chemical properties of the transition metals – the copper envelope 59
25. Reactivity of Group 2 metals 62

26.	Melting and freezing	64
27.	Diffusion in liquids	68
28.	Chemical filtration	70
29.	Rate of reaction – the effects of concentration and temperature.	73
30.	Reaction between carbon dioxide and water	76
31.	Competition for oxygen	79
32.	Making a crystal garden	83
33.	Extracting metal with charcoal	85
34.	Migration of ions	87
35.	Reduction of iron oxide by carbon	90
36.	Experiments with particles	92
37.	Particles in motion?	95
38.	Making a pH indicator	97
39.	Reaction between a metal oxide and dilute acid	99
40.	Disappearing ink	101
41.	Testing for enzymes	103
42.	Testing the hardness of water	105
43.	A chemical test for water	109
44.	Forming glass	112
45.	Thermometric titration	114
46.	Forming metal crystals	116
47.	Forming a salt which is insoluble in water	118
48.	Titration of sodium hydroxide with hydrochloric acid	120
49.	The properties of ammonia	123
50.	Causes of rusting	126
51.	Reactions of calcium carbonate	128
52.	To find the formula of hydrated copper(II) sulfate	131
53.	Heating copper(II) sulfate	134
54.	The oxidation of hydrogen	136
55.	Investigating the reactivity of aluminium	138
56.	An oscillating reaction	140
57.	Chocolate and egg	143
58.	Catalysis	145
59.	A Cartesian diver	149
60.	Neutralisation of indigestion tablets.	150
61.	Mass conservation	152
62.	Metals and acids	154
63.	Solid mixtures – a lead and tin solder	157
64.	The effect of temperature on reaction rate	159
65.	The effect of concentration on reaction rate	162
66.	The effect of heat on metal carbonates	165
67.	Change in mass when magnesium burns	169
68.	The volume of 1 mole of hydrogen gas	171

69.	How much air is used during rusting?	174
70.	Making a photographic print	176
71.	'Smarties' chromatography	179
72.	The decomposition of magnesium silicide	181
73.	An example of chemiluminescence	183
74.	Colorimetric determination of copper ore	185
75.	Glue from milk	189
76.	Rubber band	192
77.	Polymer slime	195
78.	The properties of ethanoic acid	199
79.	Properties of alcohols	201
80.	Testing salts for anions and cations	203
81.	Quantitative electrolysis	208
82.	Electrolysis of solutions	210
83.	An oxidation and reduction reaction.	213
84.	Heats of reaction (exothermic or endothermic reactions)	215
85.	Comparing the heat energy produced by combustion of various alcohols	219
86.	Fermentation	222
87.	Microbes, milk and enzymes	224
88.	The properties of the transition metals and their compounds	226
89.	Halogen compounds	230
90.	Finding the formula of an oxide of copper	233
91.	Making a fertiliser	236
92.	Electrolysis of copper(II) sulfate solution.	238
93.	Producing a foam	240
94.	Getting metals from rocks	242
95.	Addition polymerisation	245
96.	Cracking hydrocarbons	247
97.	Displacement reactions between metals and their salts	249
98.	The effect of temperature on solubility	253
99.	Purification of an impure solid	256
100.	Chemicals from seawater	258

Foreword

Chemistry is an experimental subject, and what can be more stimulating than carrying out a laboratory experiment where the results are memorable either by their visual nature or by their tying together of theory.

This collection of 100 chemistry experiments has been developed with the help and support of teachers throughout the United Kingdom. It is designed for both the experienced teacher of chemistry and for those whose first subject is not chemistry in the hope that they can communicate the excitement and wonder of the subject to their students so that they will also be captured by the subject and want to take it further.

Professor Tony Ledwith CBE PhD DSC CChem FRSC FRS
President, The Royal Society of Chemistry

Introduction

Since the introduction of the National Curriculum in England, Wales and Northern Ireland, much emphasis has been given to investigative practical work. The importance of other laboratory activities has recently become somewhat neglected. However there are many reasons for students to do class experiments.

- ▼ They allow students to apply their knowledge and understanding to what they experience, developing basic skills such as selecting and using equipment, and learning various techniques such as measuring temperature and pH.
- ▼ They often illustrate a concept or process, so students gain first-hand experience before further discussion or analysis.
- ▼ They can also be the starting point for investigations. They encourage students to ask questions and make predictions. These often arise directly from their observations.
- ▼ They extend the scope of the curriculum and can be done in chemistry or science clubs.

Furthermore, students often find experimental work stimulating, motivating and enjoyable.

The ideas for the experiments in this book have come from a variety of sources. Some may be original, but most have been collected from school chemistry departments and from the literature. Many of the ideas have come from more than one source.

Some of the experiments may be unfamiliar, others are classics. I have attempted to cover the range of experiments that would be familiar to most experienced chemistry teachers. These will be useful to new teachers and to scientists from other disciplines who are teaching chemistry, as well as to experienced subject specialists.

All of the experiments have been tested at the University of Sussex Institute of Education. They have subsequently been trialled in school where they have worked in a classroom setting. Although these experiments are reliable, teachers should try them out before using them with their students.

Kevin Hutchings

How to use this book

It is often said that when a schoolteacher writes a worksheet they reinvent the wheel; it has probably already been written elsewhere. However, most teachers would argue that their own worksheets are the most effective; they are targeted at their own classes and tailored to their own teaching style. In an attempt to address these issues, this publication allows the worksheet text and diagrams to be modified. This enables teachers to take the basic concept for the experiment, then adapt the worksheet to their own classroom needs. This should also help with another problem, where different equipment is used in the school to that stated in the resource. This normally requires the familiar explanation

'the worksheet says use a ... but we're going to use'.

From experience, this can cause exasperation for some students and confusion in less able classes. With the resource downloadable from the Internet, teachers can adapt and differentiate them as required.

There are many opportunities for customising and differentiating the student worksheets. These are really only limited by the creativity of the teacher.

For example:

▼ diagrams can be re-sized. Labels can be removed from the diagrams. Labels can be given on separate sheets. Students can then cut out and stick or copy onto their own sheet;

▼ methods can be adapted, either made into smaller steps or removed completely. This provides an opportunity for students to design their own methods. The methods can be organised into a flow diagram. This allows students to organise and sort the stages before starting the practical work. Individual steps in the procedure can be obscured so that students can then fill in the gaps. There are many possibilities for directed activities related to text (DARTs);

▼ the equipment list can be added from the teacher's guide to help students organise the apparatus; and

▼ questions can be removed, either completely or as appropriate. For example, this is common when using experiments with younger ages, when questions requiring the use of chemical formula may be too advanced.

To illustrate just a few possibilities, a number of variations of the first experiment have been included in the next section.

Adapting the worksheets – examples

Teacher's notes

The following examples show how the worksheets might be adapted.

This particular worksheet has been written for a mixed ability class of 11–12 year old students.

The examples are:

Version A
This is written in the same style as the rest of this publication.

Version B
This represents one possibility for use with less able students.

Version C
This uses unlabelled diagrams to reinforce the names of chemical apparatus.

Version D
This includes a problem solving, sequencing exercise.

Version E
This represents a more student directed approach.

Separating a sand and salt mixture

Topic

Separation techniques.

Timing

45 min.

Description

In this experiment students use simple processes to separate sand and salt.

Apparatus and equipment (per group)

- ▼ 250 cm^3 Beaker
- ▼ Filter funnel and paper
- ▼ Evaporating dish
- ▼ Tripod
- ▼ Bunsen burner
- ▼ Gauze
- ▼ Glass rod for stirring.

Chemicals (per group)

- ▼ A mixture of salt and sand (about 20 per cent salt).

Teaching tips

It can be effective to show the separate sand and salt to the whole class. Mix them at the front of the class, then use this as an introduction to a class discussion about how to separate them.

Background theory

The principles of filtration, evaporation, and the dissolving process.

Safety

Wear eye protection.

Answers

1. To dissolve the salt in water.
2. The sand is filtered out into the filter paper; the filtrate is salt solution.
3. To remove the majority of the water.

Separating a sand and salt mixture (Version A)

Introduction

In this experiment simple processes are used to separate salt from a sand and salt mixture.

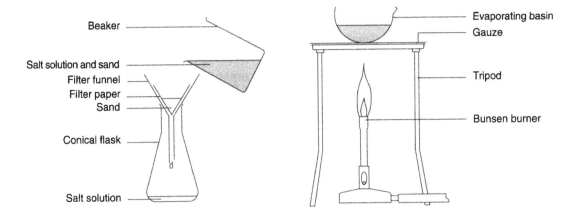

What to do

1. Mix about 5 g of the mixture with 50 cm^3 of water in a 250 cm^3 beaker. Stir gently.
2. Filter the mixture into a conical flask and pour the filtrate into an evaporating basin.
3. Heat the salt solution gently until it starts to 'spit'. **Care:** do not get too close.
4. Turn off the Bunsen burner and let the damp salt dry.

Safety

Wear eye protection.

Questions

1. Why is the salt, sand and water mixture stirred and heated in step 1?
2. What happens when this mixture is filtered in step 2?
3. Why is the salt heated in step 3?

Separating a sand and salt mixture (Version B)

Introduction

In this experiment simple processes are used to separate salt from a sand and salt mixture.

Stage 1

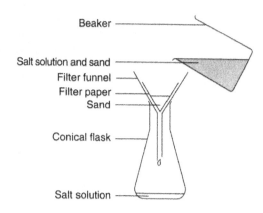

Stage 2

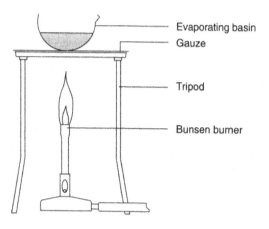

Safety

Wear eye protection.

What to do

Collect the following:

▼ 250 cm^3 Beaker
▼ Filter funnel and paper
▼ Evaporating dish
▼ Tripod

Classic chemistry experiments

- ▼ Bunsen burner
- ▼ Gauze
- ▼ Glass rod for stirring
- ▼ Salt and sand mixture (about 5 g)
- ▼ Eye protection.

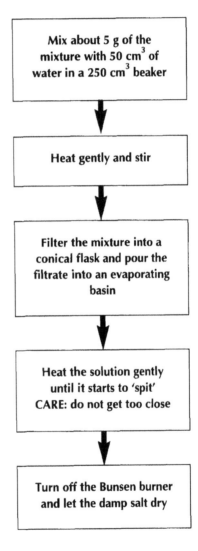

Separating a sand and salt mixture (Version C)

Introduction

In this experiment simple processes are used to separate salt from a sand and salt mixture.

Add the following labels to the diagram:

Bunsen burner, Salt solution, Conical flask, Filter funnel, Filter paper, Evaporating basin, Beaker, Sand, Salt solution and sand, Gauze, Tripod.

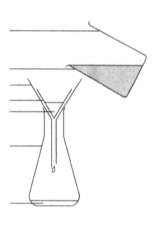 

What to do

1. Mix about 5 g of the mixture with 50 cm^3 of water in a 250 cm^3 beaker. Stir gently.
2. Filter the mixture into a conical flask and pour the filtrate into an evaporating basin.
3. Heat the salt solution gently until it starts to 'spit'. **Care:** do not get too close.
4. Turn off the Bunsen burner and let the damp salt dry.

Safety

Wear eye protection.

Questions

1. Why is the salt, sand and water mixture stirred and heated in step 1?
2. What happens when this mixture is filtered in the step 2?
3. Why is the salt heated in step 3?

Separating a sand and salt mixture (Version D)

Introduction

In this experiment simple processes are used to separate salt from a sand and salt mixture.

Add the following labels to the diagram:

Bunsen burner, Salt solution, Conical flask, Filter funnel, Filter paper, Evaporating basin, Beaker, Sand, Salt solution and sand, Gauze, Tripod.

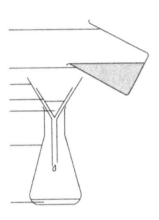 

What to do

1. Arrange these instructions into the correct order.
2. Check the accuracy with your teacher before starting the practical work.

Filter the mixture into a conical flask and pour the filtrate into an evaporating basin

Wear eye protection

Heat the solution gently until it starts to 'spit' CARE: do not get too close

Turn off the Bunsen burner and let the damp salt dry

Heat gently and stir

Mix about 5 g of the mixture with 50 cm^3 of water in a 250 cm^3 beaker

RS•C

Separation of a sand and salt mixture (Version E)

Introduction

In this experiment, simple processes will be used to separate salt from a sand and salt mixture.

Add labels to the diagram

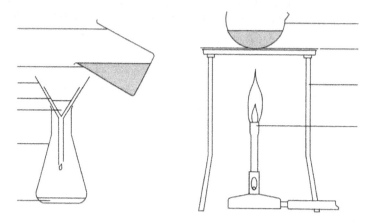

Safety

Wear eye protection.

What to do

1. Use the diagram as a guide. Complete the flow chart with a safe procedure to separate the sand and salt mixture.
2. Check with your teacher before starting the practical work.

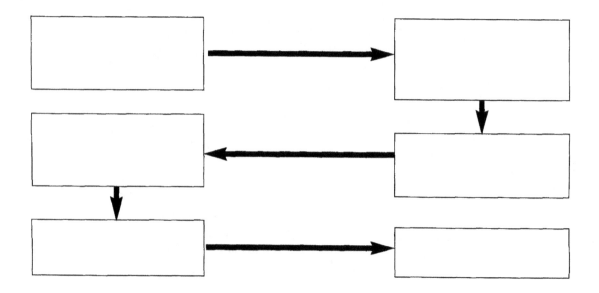

The role of information and communication technology (ICT)

Information and communication technology can be valuable in many of these classic chemistry experiments. In particular, there are sensors that measure say, temperature as a computer displays their readings on a screen. In essence the computer provides a running record of the experiment which helps students to appreciate what is happening and how fast it is happening. For example, adding a temperature sensor to the flask in an acid-base titration shows clearly that there is a point where adding further acid no longer produces heat of neutralisation. Or used in a cooling curve, a temperature sensor shows the expected temperature plateau during a change of state, while a second temperature sensor shows the change in the environment. In this latter example, students see that the environment actually gains heat while the cooling material is apparently 'doing nothing'.

Data logging technology offers some other benefits – the software usually has graph analysis tools so that rates of change can be calculated while even the smallest computer screens offer a large digital display for demonstrations. It also finds uses in longer term experiments – for example it will happily take readings overnight during the fermentation of yeast.

With or without technology, it is good practice to ask students to predict the kind of graph they expect to see.

Throughout this set of experiments, the teaching tips suggest uses of data logging sensors where, like the above examples, it can add value to the experience. Schools that have the most modest data logging system will find opportunities to use it profitably here. While few schools will have class sets of such technology, a teacher demonstration (before or after they do the experiment) can still aid students' understanding. Those with data handling software may want to use it wherever graphs need to be drawn.

Using this publication on the web

The student worksheets may be downloaded from **http://www.chemsoc.org/classic_exp** as .pdf files and as Word files that can be adapted by teachers.

Disclaimer: The Royal Society of Chemistry accepts no responsibility for any occurrence as a result of teachers modifying the worksheets. Teachers should use their professional judgement and carry out appropriate risk assessments based on the modified worksheets.

List of experiments

1. Separating a sand and salt mixture
2. Viscosity
3. Rate of evaporation
4. Chromatography of leaves
5. The energetics of freezing
6. Accumulator
7. Electricity from chemicals
8. Iron in breakfast cereal
9. Unsaturation in fats and oils
10. The pH scale
11. Preparation and properties of oxygen
12. Identifying polymers
13. Energy values of food
14. A compound from two elements
15. Chemistry and electricity
16. Combustion
17. Determining relative atomic mass
18. Reaction of a Group 7 element (iodine with zinc)
19. Reactions of halogens
20. Sublimation of air freshener
21. Testing the pH of oxides
22. Exothermic or endothermic?
23. Water expands when it freezes
24. Chemical properties of the transition metals - the copper envelope
25. Reactivity of Group 2 metals
26. Melting and freezing
27. Diffusion in liquids
28. Chemical filtration
29. Rate of reaction – the effects of concentration and temperature
30. Reaction between carbon dioxide and water
31. Competition for oxygen
32. Making a crystal garden
33. Extracting metal with charcoal
34. Migration of ions
35. Reduction of iron oxide by carbon
36. Experiments with particles
37. Particles in motion?
38. Making a pH indicator
39. Reaction between a metal oxide and dilute acid
40. Disappearing ink
41. Testing for enzymes
42. Testing the hardness of water
43. A chemical test for water
44. Forming glass
45. Thermometric titration
46. Forming metal crystals
47. Forming a salt which is insoluble in water
48. Titrating sodium hydroxide with hydrochloric acid
49. The properties of ammonia
50. Causes of rusting
51. Reactions of calcium carbonate
52. To find the formula of hydrated copper(II) sulfate

53. Heating copper(II) sulfate
54. The oxidation of hydrogen
55. Investigating the reactivity of aluminium
56. An oscillating reaction
57. Chocolate and egg
58. Catalysis
59. Cartesian diver
60. Neutralisation of indigestion tablets
61. Mass conservation
62. Metals and acids
63. Solid mixtures – a lead and tin solder
64. The effect of temperature on reaction rate
65. The effect of concentration on reaction rate
66. The effect of heat on metal carbonates
67. Change in mass when magnesium burns
68. The volume of 1 mole of hydrogen gas
69. How much air is used during rusting?
70. Making a photographic print
71. 'Smarties' chromatography
72. The decomposition of magnesium silicide
73. An example of chemiluminescence
74. Colorimetric determination of copper ore
75. Glue from milk
76. Rubber band
77. Polymer slime
78. The properties of ethanoic acid
79. The properties of alcohols
80. Testing salts for anions and cations
81. Quantitative electrolysis
82. Electrolysis of solutions
83. An oxidation and reduction reaction
84. Heats of reaction (exothermic or endothermic reactions)
85. Comparing the heat energy produced by combustion of various alcohols
86. Fermentation
87. Microbes, milk and enzymes
88. The properties of the transition metals and their compounds
89. Halogen compounds
90. Finding the formula of an oxide of copper
91. Making a fertiliser
92. Electrolysis of copper(II) sulfate solution
93. Producing a foam
94. Getting metals from rocks
95. Addition polymerisation
96. Cracking hydrocarbons
97. Displacement reactions between metals and their salts
98. The effect of temperature on solubility
99. Purification of an impure solid
100. Chemicals from seawater

List of experiments by categories

Here the experiments are listed by categories. There are also suggestions for the place of each experiment within the curriculum. These should not be taken as prescriptive. There are very few experiments that could not cross this somewhat artificial pre-14 and post-14 boundary. Teachers can adapt the experiments according to the needs of their students and there are many that can be revisited at various times within the curriculum.

Experiment No	1	2	3	4	5	6	7	8	9	10	11	12	13	14	15	16	17	18	19	20
Curriculum Pre-14	✓	✓	✓	✓	✓		✓	✓		✓	✓	✓	✓	✓	✓	✓				✓
Post-14				✓	✓	✓	✓		✓	✓		✓	✓		✓		✓	✓	✓	✓
Periodicity																				
Acids and bases	✓																			
Physical properties		✓	✓		✓	✓	✓			✓					✓		✓	✓	✓	
Particulate nature of matter		✓	✓	✓	✓			✓	✓		✓	✓		✓	✓			✓	✓	✓
Chemical analysis				✓	✓										✓				✓	✓
Formation of compounds				✓								✓	✓	✓		✓				
Mixtures and separations				✓	✓	✓		✓				✓	✓	✓	✓					
Energy	✓						✓						✓			✓				✓
Organic chemistry									✓			✓				✓				
Rates of reaction																				

Experiment No	21	22	23	24	25	26	27	28	29	30	31	32	33	34	35	36	37	38	39	40
Curriculum Pre-14	✓	✓	✓	✓	✓		✓	✓		✓	✓	✓	✓		✓	✓	✓			✓
Post-14	✓	✓		✓	✓	✓	✓	✓	✓	✓			✓	✓	✓		✓	✓	✓	
Periodicity	✓				✓															
Acids and bases												✓								
Physical properties			✓	✓						✓		✓						✓	✓	✓
Particulate nature of matter			✓			✓	✓	✓	✓	✓	✓			✓		✓				
Chemical analysis					✓	✓	✓			✓				✓			✓	✓		
Formation of compounds				✓						✓			✓	✓	✓		✓		✓	
Mixtures and separations		✓	✓											✓						
Energy								✓						✓						
Organic chemistry						✓														
Rates of reaction									✓											

RS·C

Classic chemistry experiments

Experiment No	41	42	43	44	45	46	47	48	49	50	51	52	53	54	55	56	57	58	59	60
Curriculum Pre-14	✓	✓	✓	✓	✓	✓	✓			✓	✓		✓	✓	✓	✓	✓		✓	
Post-14	✓	✓	✓	✓	✓	✓	✓	✓	✓			✓	✓	✓		✓		✓		
Periodicity																		✓		
Acids and bases								✓										✓		
Physical properties	✓				✓						✓				✓					
Particulate nature of matter	✓										✓									
Chemical analysis		✓	✓	✓		✓		✓	✓			✓	✓			✓	✓		✓	✓
Formation of compounds		✓		✓	✓	✓	✓		✓	✓		✓	✓	✓			✓			✓
Mixtures and separations					✓			✓	✓			✓								
Energy					✓				✓					✓	✓					
Organic chemistry																				
Rates of reaction														✓	✓	✓		✓		

Experiment No	61	62	63	64	65	66	67	68	69	70	71	72	73	74	75	76	77	78	79	80
Curriculum Pre-14	✓	✓	✓	✓	✓	✓	✓	✓	✓	✓	✓	✓	✓	✓	✓	✓	✓	✓	✓	✓
Post-14		✓	✓	✓	✓	✓		✓		✓	✓	✓	✓	✓		✓	✓	✓	✓	✓
Periodicity																				
Acids and bases	✓	✓	✓	✓	✓	✓		✓									✓			
Physical properties	✓																			
Particulate nature of matter							✓	✓	✓	✓	✓	✓	✓	✓	✓	✓	✓	✓	✓	✓
Chemical analysis			✓	✓	✓		✓	✓	✓					✓	✓	✓	✓			
Formation of compounds										✓	✓	✓	✓	✓	✓					
Mixtures and separations														✓						
Energy										✓		✓	✓		✓		✓	✓		
Organic chemistry																✓			✓	
Rates of reaction				✓		✓												✓	✓	

Classic chemistry experiments

Experiment No	81	82	83	84	85	86	87	88	89	90	91	92	93	94	95	96	97	98	99	100
Curriculum Pre-14	✓	✓	✓				✓	✓	✓	✓		✓	✓	✓	✓	✓		✓	✓	✓
Post-14		✓	✓	✓	✓	✓	✓	✓	✓		✓			✓			✓	✓		
Periodicity																				
Acids and bases								✓	✓											
Physical properties	✓	✓						✓	✓		✓									
Particulate nature of matter	✓	✓	✓									✓	✓	✓				✓	✓	✓
Chemical analysis	✓							✓	✓	✓	✓	✓	✓				✓	✓	✓	✓
Formation of compounds		✓	✓	✓		✓	✓	✓	✓	✓				✓						
Mixtures and separations				✓										✓						
Energy	✓																		✓	✓
Organic chemistry					✓	✓	✓								✓	✓				
Rates of reaction					✓		✓													

RS·C

Health and safety

The purpose of this book is to give examples of good practice in hands-on chemistry teaching. We believe that all the activities can be carried out safely in schools. The hazards have been identified and any risks from them reduced to insignificant levels by the adoption of suitable control measures. However, we also think it is worth explaining the strategies we have adopted to reduce the risks in this way.

Regulations[1] made under the Health and Safety at Work *etc* Act 1974 require a risk assessment to be carried out before hazardous chemicals are used or made, or a hazardous procedure is carried out. Risk assessment is your employer's responsibility. The task of assessing risk in particular situations may well be delegated by the employer to the head of science/chemistry, who will be expected to operate within the employer's guidelines. Following guidance from the Health and Safety Executive most education employers have adopted various nationally available texts as the basis for their model risk assessments. Those commonly used include the following:

Safeguards in the School Laboratory, 10th edition, ASE, 1996
Topics in Safety, 2nd edition, ASE, 1988
Hazcards, CLEAPSS[2], 1998 (or 1995)
Laboratory Handbook, CLEAPSS[2], 1997
Safety in Science Education, DfEE, HMSO, 1996
Hazardous Chemicals Manual, SSERC[2], 1997.

If your employer has adopted one or more of these publications, you should follow the guidance given there, subject only to a need to check and consider whether minor modification is needed to deal with the special situation in your class/school. We believe that all the activities in this book are compatible with the model risk assessments listed above. However, teachers must still verify that what is proposed does conform with any code of practice produced by their employer. You also need to consider your local circumstances. Is your fume cupboard reliable? Are your students reliable?

Risk assessment involves answering two questions:

- how likely is it that something will go wrong? and
- how serious would it be if it did go wrong?

Hydrogen has been exploding (see Experiment 54) ever since people started teaching chemistry – and long may it continue to do so! But the explosions must be carried out in a controlled way, so as to avoid injury, and unintended explosions must be avoided. You need to be reasonably sure that no student will attempt to light the hydrogen at the delivery tube, because there is a risk that the 'hydrogen' is in fact an explosive mixture of hydrogen and air.

How likely it is that something will go wrong depends on who is doing it and what sort of training and experience they have had. You would obviously not ask 11 year old students to heat concentrated sulfuric acid with sodium bromide, because their inexperience and lack of practical skills makes a serious accident all too likely. By the time they reach post-16 they should have acquired the skills and maturity to carry it out safely. In most of the publications listed above there are suggestions as to whether an activity should be a teacher demonstration only, or could be done by students of various ages. Your employer will probably expect you to follow this guidance. This means, for example, that the Addition polymerisation (Experiment 95) should normally only be done as a teacher demonstration or by post-16 students. Perhaps with well-motivated and able students it might be done pre-16, but any deviation from the model risk assessment needs discussion and a written justification beforehand. More commonly, teachers will conclude that an activity which is, in principle, permissible is really not suitable for their students.

Teachers tend to think of eye protection as the main control measure to prevent injury. In fact, personal protective equipment, such as goggles or safety spectacles, is meant to protect from the unexpected. If you expect a problem, more stringent controls are needed. A range of control measures may be adopted, the following being the most common. Use:

- a less hazardous (substitute) chemical;
- as small a quantity as possible;
- as low a concentration as possible;
- a fume cupboard; and
- safety screens (more than one is usually needed, to protect both teacher and students).

The importance of lower concentrations is not always appreciated, but the following table, showing the hazard classification of a range of common solutions, should make the point.

Chemical	Lower hazard	Higher hazard
ammonia (aqueous)	irritant if ≥ 3 M	corrosive if ≥ 6 M
sodium hydroxide	irritant if ≥ 0.05 M	corrosive if ≥ 0.5 M
ethanoic (acetic) acid	irritant if ≥ 1.5 M	corrosive if ≥ 4 M
hydrochloric acid	irritant if ≥ 2 M	corrosive if ≥ 6.5 M
nitric acid	irritant if ≥ 0.1 M	corrosive if ≥ 0.5 M
sulfuric acid	irritant if ≥ 0.5 M	corrosive if ≥ 1.5 M
barium chloride	hamrful if ≥ 0.02 M	toxic if ≥ 0.2 M (or if solid)
copper(II) sulfate	harmful if ≥ 1 M (or if solid)	
copper(II) chloride	harmful if ≥ 0.15 M	toxic if ≥ 1.4 M (or if solid)
iron(II) sulfate	harmful if ≥ 1 M	
iron(III) sulfate	harmful if ≥ 0.75 M	
lead nitrate	harmful if ≥ 0.001 M	toxic if ≥ 0.01 M
potassium dichromate	toxic if ≥ 0.003 M	very toxic if ≥ 0.2 M (or if solid)
silver nitrate	irritant if ≥ 0.2 M	corrosive if ≥ 0.5 M (or if solid)
bromine water	harmful & irritant if ≥ 0.0006 M (0.1 per cent) toxic & corrosive if ≥ 0.06 (1 per cent) very toxic & corrosive if ≥ 0.42 M (7 per cent) (non aqueous or saturated aqueous solution)	

Throughout this book, we make frequent reference to the need to wear eye protection. Undoubtedly, chemical splash goggles, to the European Standard EN 166 3 give the best protection but students are often reluctant to wear goggles. Safety spectacles give less protection, but may be adequate if nothing which is classed as corrosive or toxic is in use. Reference to the above table will show, therefore, that if sodium hydroxide is in use, it should be more dilute than 0.5 M.

Science as carried out in schools is very safe. Do not believe the myths and rumours that various chemcials or procedures are banned. That is very rarely true, although you may well have to take various safety precautions. Chemistry can be – and should be – fun. It must also be safe. The two are not incompatible.

Dr Peter Borrows MA PhD CChem FRSC
Director, CLEAPSS School Science Service

[1] The COSHH Regulations and the Management of Health and Safety at Work Regulations.
[2] Note that CLEAPSS and SSERC publications are available only to members.
CLEAPSS School Science Service, Brunel University, Uxbridge UB8 3PH
Tel: 01895 251496   Fax: 01895 814372   Email: science@cleapss.org.uk
SSERC, St Mary's Building, 23 Holyrood Rd, Edinburgh EH8 8AE
Tel: 0131 558 8180   Fax: 0131 558 8191   Email: sserc@mhic.ac.uk

Acknowledgements

The production of this book was only made possible because of the advice and assistance of a large number of people. To the following, and to everybody who has been involved with this project, both the author and the Royal Society of Chemistry express their gratitude.

M. Anstiss	King's Manor School, West Sussex
E. Barker	St Mary's School, Colchester
S. Barrett and P. Starbuck	Ashby Grammar School, Ashby-de-la-Zouch, Leicestershire
P. Bartle	Institute of Education, University of Sussex
P. Battye	York Sixth Form College, York
M. Berry	Chislehurst & Sidcup GS, Kent
Mr Blow	Yarborough School, Lincoln
I. Boundy	Institute of Education, University of Sussex
G.P. Bridger	Bideford College, Bideford
C. Buck	Northampton School for Boys, Northampton
G. Burleigh	Nailsea School, North Somerset
A. Campbell	Downend School, Bristol
D. Carr	Caistor Grammar School, Caistor
M. J. Carr	Sir Thomas Rich's, Gloucester
A. Cattawach	Stevenson College, Edinburgh
L. Campbell	Greencroft Comprehensive School, Stanley, Co Durham
P. Calder	Leek High School, Leek, Staffordshire
H. Champness	Heanor Gate School, Heanor
Chemistry Department	Aston Comprehensive School, Aston
Chemistry Department	Nottingham High School, Nottingham
D.A. Cooper	Sutton Valence School, Kent
G. Collier	Dollar Academy, Dollar
E. Convery	Institute of Education, University of Sussex
R. Couteur	King Edward VI School, Chelmsford
J. Craig	Groby Community College, Groby
D.R. Crilley	St Leonard's RC Comprehensive, Durham
J. Cross	Lordswood Girl's School, Birmingham
J. Davies	Hipperholme & Lightcliffe High School, Halifax
L. Drew	St Catherine's School, Guildford
H. Dunlop	American Community School, Surrey
C. Edwards	Sutton Community High School, Merseyside
M. Evetts	Burford School, Burford
H. Fletcher and M. Bell	Tonyrefail School, Tonyrefail
C.A. Freeborn	Birchgrove Comprehensive School, Swansea
J. Gillin	Newquay Tueriglas, Cornwall
K. Grabowski	All Saint's RC School, Mansfield, Nottinghamshire
D. Gunputh	Dorothy Stringer School, Brighton
C. Gill	Theak Green Community School, Reading
D. Groat	Thurso High School, Caithness
T. Harrison	Rednock School, Gloucestershire
A.D. Heaton	Coquet High School, Northumberland
B. Hey	Dixons CTC, Bradford
R. Higgins	Framlingham College, Suffolk
D.M.R. Hill	Vermuyden School, East Yorkshire
J. Hind	Reading Girls School, Reading
K. Holsgrove	Weaverham High School, Cheshire
M. Howson	The Portsmouth Grammar School, Portsmouth
N.J. Hoyle	St Helen's School, Middlesex

N. Huddleston	Holy Trinity School, Crawley
M. Hudson	Chesterfield High School, Liverpool
T. Hughes	Colne Community School, Essex
A. Hutchinson	Rye Hills School, Redcar
C. Hutchinson	Fairfield High School, Widnes, Cheshire
N. Jackson	Belmont Comprehensive School, Durham
P. James	Mathematical School, Rochester
A. Jenkins	Bishop Vaughan RC Comprehensive School, Clase Morriston, Swansea
J. Johnson	Heanor Gate School, Derby
J. Kember	Institute of Education, University of Sussex
S. Kendall	John Kitto Community College, Devon
M. Khalid	The Matthew Holland School, Nottingham
J. King	Institute of Education, University of Sussex
D. Knight	Aldernbrook School, Solihull
D. Leese	Wentworth High School, Eccles
E.J. Lewis	Crambrook School, Kent
S. Linwood	Institute of Education, University of Sussex
S. Logan	Huddersfield Road, Oldham
H.S. MacDonald	Sevenoaks School, Kent
J. McAdam	Geary Drive, Essex
M.S. McLaren	Granton Grammar School, Grantown on Spey
J. Machin	Shalstone Road, London
T. Martin	Saffron Walden County High School, Saffron Walden, Essex
T. Meunier	Dragon School, Oxford
A. Miller	Seaford Head Community College, East Sussex
J. Mitenree	Sir E Scott School, Isle of Harris
D.S. Moore	St Edward's School, Oxford
S. Morris	The Cotswold School, Cheltenham
G. Nethercroft	Enfield County School, Enfield
C. Newman	Bedlingtonshire Community High School, Bedlington
G. Newman	Institute of Education, University of Sussex
T. Nicolas	Institute of Education, University of Sussex
M. Oakes	Chesterton High School, Staffordshire
J. Oldfield	Leeds Girls High School, Leeds
F. O'Farrell	International School of Dusseldorf, Dusseldorf
B. Orger	Stone School, Buckingham
P. Oddy	Mullion School, Cornwall
E.X.Orry	Rochester Grammar School, Rochester
T. Packer	Longcroft School, Beverley, East Riding
J.S. Page	Clevedon School, Clevedon, North Somerset
A. Pickles	Blessed Robert Johnson Catholic College, Shropshire
W. Pitt	King's School, Cambridgeshire
E. Poole	Roade School, Northamptonshire
T.R.Read	Finchley Catholic High School, North Finchley, London
R. Reynolds	St Louis Grammar School, Kilkeel
G. Riley	King Edward VI Grammar School, Chelmsford
E. Richardson	Bedlinetonshire High School, Northumberland
G. Rishman	Chiselhurst and Sidcup Grammar School, Kent
V. Rigby	Padgate High School, Warrington
G. Riley	King Edward Grammar School, Essex
J. Roberts	Leasowes High School and Community College, West Midlands
J. Rogers	Charters School, Berkshire
S. Rust	Maiden Erleigh School, Reading

S. Rutland	Wellington Crescent, Norfolk
K. Ryan	Cherwell School, Oxford
T. Saleh	Croxteth Community Comprehensive, Liverpool
J.P. Seaman	Maidstone Grammar School, Kent
M.S. Shauq	Blekeston School, Stockton on Tees
D. Shaw	The Grange School, Hartford, Cheshire
J. Shields	Barrow Sixth Form College, Cumbria
Mr K.A. Simpson	QECC, Crediton
S. Sims	Lever Brothers Ltd, Port Sunlight
J. Slater	Dalriada School, Co Antrim
C. Smith	Archbishop Sarcroft High School, Harleston
K. Smith	Queens School, Wisbech
J.P. Smith	Cavendish School, Eastbourne
M.J. Snell	Thornhill School, Thornholme Road, Tyne and Wear
L. Stanbury	St Albans School, St Albans
P. Stanley	Regis School, Wolverhampton
J. Starling	Sackville School, East Grinstead
A. Story	Mathematical School, Rochester
R.J. Suffolk	Chichester High School For Girls, Chichester West Sussex
D. Swales	The Kingstone School, South Yorkshire
A.Z. Szydlo	Highgate School, London
D. Taylor	Calderstones CCS, Liverpool
R. Taylor	The City Technology College, Kingshurst
J. Taylor	Great Cornard Upper School, Sudbury
P. Thacker	Highfields School, Derbyshire
G.A. Thomas	Llantwit Major School, S. Glam
A. Thorpe	Ampleforth College, York
D.M. Titchener	Ysgol Bryn Offer, Wrexham
A. Tuff	Alec Hunter High School, Essex
G. Vergara	The Grange Community School, Warmley, Bristol
E. Vicars	Greenacre School, Surrey
J. Wainwright	Colyton Grammar School, Devon
D. Waistnidge	King Edward VI College, Devon
A. Waller	Saltash Community School, Cornwall
S.N. Whittleton	Christ the King School, Southport
B. Wild	Cheltenham College, Cheltenham
G. Woods	Sycamores Westfield Road, Gwent
J. Wray	Stoke Newington School, London
R.A. Wright	Charterhouse, Surrey
E. Yendell	Institute of Education, University of Sussex
D.R. Young	Deacons School, Peterborough

The Royal Society of Chemistry thanks the Chemical Industry Education Centre at the University of York for allowing the inclusion of Experiment No 12: Identifying Polymers. This is adapted from *Recycling Cities*.

The Society also thanks Peter Borrows of CLEAPSS School Science Service, Brunel University, and Chair of the Association for Science Education Safeguards in Science Committee, for advice on safety.

The Society also thanks Roger Frost of IT in Science for advice on the use of ICT.

The Society would like to extend its gratitude to the University of Sussex Institute of Education for providing laboratory and office accommodation for this Fellowship, and the Head Teacher and Governors of Seaford Head Community College, Seaford, East Sussex for seconding Kevin Hutchings to the Society's Education Department.

Finally, the author thanks the staff of the education department at the Royal Society of Chemistry, especially Colin Osborne and John Johnston, for their advice, encouragement and unfailing support throughout the project.

Bibliography

1. Shakhashiri, *Chemical Demonstrations*, **1–4**. Madison Wisconsin: University of Wisconsin Press, 1982
2. Summerlin et al, *Chemical Demonstrations*, **1 & 2**. Washington: American Chemical Society, 1985.
3. Borgford & Summerlin, *Chemical activities*. Washington: American Chemical Society, 1988.
4. Experiments from School Science Review, Science Masters Book series **1–4**, London: John Murray 1931–64
5. Carl H Snyder, *The Extraordinary Chemistry of Ordinary Things*. Chichester: John Wiley, 1992.
6. Liptrot and Pode, *Exploring Chemistry*. Richmond: Mills & Boon, 1968.
7. Barker & Knapp, *Chemistry – a Practical Approach*. Macmillan Education, 1978.
8. T. Hilton, *Recycling Cities Making use of Science and Technology*. York: Chemical Industry Education Centre, 1992.
9. G. Snape et al, *Science at work*, 14-16. Harlow: Addison Wesley Longman Limited, 1996
10. *Oooh Ahhh! Awesome Experiments!* Colorado Springs: Steve Spangler Current Inc, 1994.
11. E. Ramsden, *Key Science: Chemistry Extension File* Cheltenham: Stanley Thornes Publishers, 1997
12. E.T. Richardson, *Science Now!* Oxford: Heinemann Educational, 1996.
13. Jean McLean, *Chemistry*. Harlow: Addison Wesley Longman Limited, 1996.
14. Hollins et al, *Go for Science!* Walton on Thames: Thomas Nelson & Sons Ltd, 1998.
15. Bethell et al, *Co-ordinated science Teachers Guide Chemistry*. Oxford: Oxford University Press, 1996
16. Gott et al, *Science Investigations*. London: Collins Educational, 1997.
17. J. Jones (ed), *Science Plus*. London: Collins Educational, 1996.
18. DFEE, *Safety in Science Education*. London: Stationery Office, 1996.
19. Lawrie Ryan, *Chemistry for You*. Cheltenham: Stanley Thornes, 1996
20. R Gallagher and P Ingram, *Chemistry Made Clear*. Oxford: Oxford University Press, 1984
21. T. Lister, *Classic Chemistry Demonstrations*. London: Royal Society of Chemistry Educational Division, 1995
22. J. Skinner, *Microscale Chemistry*. London: Royal Society of Chemistry Educational Division, 1997.
23. Nuffield Chelsea Curriculum Trust, *Chemical Activities for GCSE*. London: Nuffield, 1996.
24. T. Parkyn, *Chemistry Copymasters*. Harlow: Longman, 1996.
25. *Revised Nuffield Chemistry Books I, II & III*, Harlow: Longman, 1978.
26. John Holman, *Chemistry*. Walton on Thames: Nelson, 1995.
27. S.C. Rust, School Science Review, 1988, 70(**250**), 73–75
28. M.H. Gabb and W.E. Latchchem, *A Handbook of Laboratory Solutions*. London Andre Deutsch, 1967.

The Experiments

RS•C

1. Separating a sand and salt mixture

Topic

Separation techniques.

Timing

45 min.

Description

In this experiment students use simple processes to separate sand and salt.

Apparatus and equipment (per group)

- 250 cm^3 Beaker
- Filter funnel and paper
- Evaporating dish
- Tripod
- Bunsen burner
- Gauze
- Glass rod for stirring.

Chemicals (per group)

A mixture of salt and sand (about 20 per cent salt).

Teaching tips

It can be effective to show the separate sand and salt to the whole class. Mix them at the front of the class, then use this as an introduction to a class discussion about how to separate them.

Background theory

The principles of filtration, evaporation, and the dissolving process.

Safety

Wear eye protection.

Answers

1. To dissolve the salt in water.
2. The sand is filtered out into the filter paper; the filtrate is salt solution.
3. To remove the majority of the water.

Separating a sand and salt mixture

Introduction

In this experiment simple processes are used to separate salt from a sand and salt mixture.

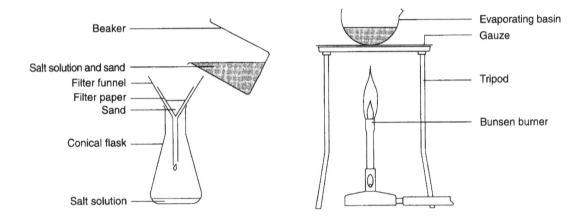

What to do

1. Mix about 5 g of the mixture with 50 cm^3 of water in a 250 cm^3 beaker. Stir gently.
2. Filter the mixture into a conical flask and pour the filtrate into an evaporating basin.
3. Heat the salt solution gently until it starts to 'spit'. **Care:** do not get too close.
4. Turn off the Bunsen burner and let the damp salt dry.

Safety

Wear eye protection.

Questions

1. Why is the salt, sand and water mixture stirred in step 1?
2. What happens when this mixture is filtered in the step 2?
3. Why is the salt heated in step 3?

2. Viscosity

Topic

Viscosity.

Timing

20 min.

Description

Students are provided with a set of identical tubes each containing a different liquid. Students measure the time taken for a bubble to rise through the liquid. This is used to compare the viscosity of the liquids.

Apparatus and equipment (per group)

▼ Stopclock.

Chemicals (per group)

▼ Prepared sealed tubes of different liquids

The plastic tubes in which thermometers are packaged are ideal for this purpose. Seal one end of the tube using super-glue and the plastic stopper. Fill the tube with liquid leaving a measured gap of 1 cm. Seal the other end in the same way. Check for leaks before giving the tube to your students.

Liquids to choose could include:

Water, cooking oil, washing up liquid, fresh engine or clutch oil (NOT used oils), ethanol (**Highly flammable**), propan-1,2,3-triol (glycerol), shampoo or bubble bath.

Teaching tips

Remind students to time each liquid using a consistent method – eg measure the time from inversion until the 'bubble first hits the top'.

Background theory

The particulate nature of matter.

Viscosity

Introduction

The viscosity of a liquid is another term for the thickness of a liquid. Thick treacle-like liquids are viscous, runny liquids like water are less viscous. Gases exhibit viscosity in the same way. In this experiment, the viscosity of various liquids are compared.

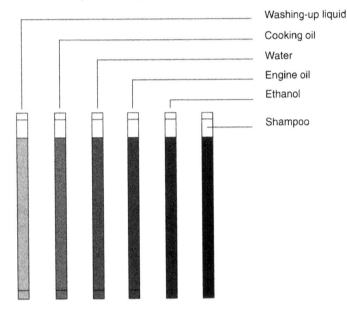

What to record

Complete a table like this:

Liquid	Time taken /s
Washing up liquid	
Water	

What to do

1. Take one of the tubes provided.
2. Ensure the bubble is at the top and the tube is held vertical.
3. Quickly invert the tube and measure the time it takes for the bubble to reach the top.
4. Repeat this measurement for all the samples.

Questions

1. Which liquid is the most viscous?
2. Which liquid is the least viscous?
3. Design a different experiment for comparing the viscosity of liquids.

3. Rate of evaporation

Topic

Evaporation, kinetic theory.

Timing

50 min.

Description

Students place a single drop of propanone on a microscope slide and measure the time for the drop to evaporate. Different conditions are tested, and compared.

Apparatus and equipment (per group)

- ▼ Microscope slide
- ▼ Stopclock
- ▼ Used match stick.

Chemicals (per group)

Propanone (**Highly flammable**) bottles with droppers (preferably all the same size droppers to ensure comparable results). Use the type of teat pipettes (usually fitted to Universal Indicator bottles) that do not allow squirting
– eg Griffin.

Teaching tips

Propanone must be dry, slides can be warmed in water and quickly dried for temperature variation.

Temperature sensors attached to a computer can be used as an adjunct to this experiment. They provide another way of quantifying the rate of evaporation and you might demonstrate this approach at the end of a lesson or use the idea as a basis for a separate investigation. The propanone is dropped onto temperature probes as a graph is displayed on the screen. One of the probes can be placed in an air flow and the graph slope will provide a measure of the rate of evaporation. Ask the students to predict the effect of using a warm flow over a volatile liquid – the temperature may drop before it rises.

Background theory

Kinetic theory.

Safety

Ensure there are no sources of ignition nearby. Wear eye protection.

Answers

1. Temperature, surface area, extent of air movement.
2. Since faster particles escape from the surface of the liquid during evaporation, the net kinetic energy of the remaining particles is lowered, therefore evaporation causes cooling.

Rate of evaporation

Introduction

Evaporation is the conversion of a liquid into vapour, without necessarily reaching the boiling point. In this experiment the rate of evaporation is measured and compared under various different conditions.

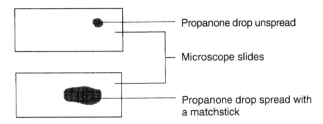

- Propanone drop unspread
- Microscope slides
- Propanone drop spread with a matchstick

What to record

Complete the following table.

Condition	Evaporation time (s)
Unspread, cool, air movement	
Unspread, cool, no air movement	
Spread out, cool, no air movement	
Spread out, warm, no air movement	
Unspread, warm, air movement	
Spread out, cool, air movement	
Spread out, warm, air movement	
Unspread, warm, no air movement	

What to do

1. Consider the following conditions for the evaporation of a drop of propanone on a microscope slide.

Condition	How achieved
Warm	Warm slide in hands and hold on a flat palm. Alternatively, place the slide in warm water then dry the slide.
Cool	Room temperature.
Spread out drop	Spread the drop of propanone on the slide with a matchstick.
Unspread	Drop left as one drop on the slide.
Cool air flow	Fan with book.
Warm air flow	Blow across drop.

2. Place a microscope slide in one of the conditions listed.
3. Add the single drop of propanone.
4. Measure the time for the drop to evaporate.
5. Repeat the experiment using different conditions.

Safety

Ensure there are no sources of ignition nearby. Wear eye protection.

Questions

1. Name three factors that affect the rate of evaporation.
2. Why does evaporation produce a cooling effect?

4. Chromatography of leaves

Topic

Separation of mixtures, extraction.

Timing

30 min.

Description

Students use chromatography to separate the pigments present in a leaf.

Apparatus and equipment (per group)

- ▼ Mortar and pestle
- ▼ Chromatography paper
- ▼ Teat pipette. Use the type of teat pipettes (usually fitted to Universal Indicator bottles) that do not allow squirting – *eg* Griffin.
- ▼ 100 cm^3 Beaker
- ▼ Small capillary tube to transfer drops of liquid onto chromatography paper.

Chemicals (per group)

Propanone (**Highly flammable**)

Sand

Cut-up leaves or leaves and scissors.

Teaching tips

This works well if a very concentrated solution is prepared. Use cut up leaves, a pinch of sand, and a few drops of solvent. (Some separation may occur at stage 6).

Background theory

Different substances have different attractions to the paper.

Safety

Wear eye protection. Avoid naked flames.

Answers

1. Two.
2. Green and yellow.
3. Yellow moved further than green.

Chromatography of leaves

Introduction

Most leaves are green due to chlorophyll. This substance is important in photosynthesis (the process by which plants make their food). In this experiment, the different pigments present in a leaf are separated using paper chromatography.

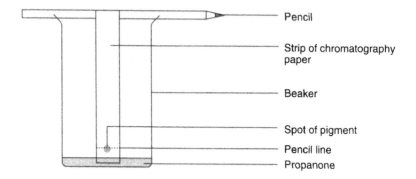

What to record

The chromatogram produced in this experiment can be dried and kept.

What to do

1. Finely cut up some leaves and fill a mortar to about 2 cm depth.
2. Add a pinch of sand and six drops of propanone from the teat pipette.
3. Grind the mixture for at least three minutes.
4. On a strip of chromatography paper, draw a pencil line 3 cm from the bottom.
5. Use a fine glass tube to put liquid from the leaf extract onto the centre of the line. Keep the spot as small as possible.
6. Allow the spot to dry, then add another spot on top. Add five more drops of solution, letting each one dry before putting on the next. The idea is to build up a very concentrated small spot on the paper.
7. Put a small amount of propanone in a beaker and hang the paper so it dips in the propanone. Ensure the propanone level is below the spot.
8. Leave until the propanone has soaked near to the top.
9. Mark how high the propanone gets on the paper with a pencil and let the chromatogram dry.

Safety

Propanone is highly flammable. Wear eye protection.

Questions

1. How many substances are on the chromatogram?
2. What colours are they?
3. Which colour moved furthest?

Classic chemistry experiments

RS•C

5. The energetics of freezing

Topic

Particle theory, states of matter, energy changes.

Timing

15 min.

Description

Students melt sodium thiosulfate crystals, these are cooled to below the melting point. They exist in a metastable supercooled state. The supercooled liquid will freeze on the addition of a crystal of sodium thiosulfate, or dust particles or on stirring. Students stir the supercooled liquid with a thermometer and observe the temperature increase.

Apparatus and equipment (per group)

- ▼ Test-tube
- ▼ Stirring thermometer (-10–110 °C)
- ▼ Bunsen burner
- ▼ Test-tube holder
- ▼ Cotton wool tuft
- ▼ 100 cm^3 Beaker.

Chemicals (per group)

Sodium thiosulfate pentahydrate ($Na_2S_2O_3.5H_2O$).

Teaching tips

It may be better to stand the test-tube in a hot water bath to melt the solid. Take care not to break the thermometer in solidified sodium thiosulfate. It is probably best to let it solidify. Sodium thiosulfate is soluble and a thermometer can easily be removed by flushing with water.

A temperature sensor attached to a computer can be used in place of a thermometer. It can plot the temperature change on a graph and show this as it occurs. It should show a temperature plateau as the liquid turns solid. However, a slight modification of the experiment can yield an intriguing result. Place the cooling test-tube in an insulated cup containing a few cm^3 of water. Use a second temperature sensor to monitor the temperature of the water. The temperature should rise as the thiosulfate cools and it should continue to rise even as the thiosulfate changes state.

Background theory

The particle model for solids, liquids and gases.

Safety

Wear eye protection.

Answers

1. Exothermic.
2. Room temperature is lower than the melting point of sodium thiosulfate pentahydrate, the sodium thiosulfate pentahydrate cools to this temperature.

Energetics of freezing

Introduction

When a substance changes state, energy can be produced or absorbed. This experiment illustrates the energy change when a liquid freezes to form a solid.

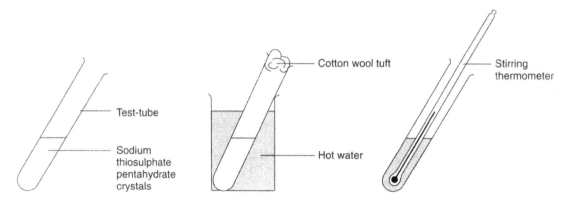

What to record

Record the temperature of the liquid, record the temperature as the liquid solidifies (this is the melting point of sodium thiosulfate pentahydrate).

What to do

1. Half fill a test-tube with crystals of sodium thiosulfate pentahydrate.
2. Warm the test-tube gently in a beaker of hot water to melt the crystals.
3. Put a tuft of cotton wool in the top of the test-tube to exclude dust.
4. Stand the test-tube in an empty beaker and leave in a still place to cool.
5. Remove the cotton wool, put a thermometer in the melt, and record the temperature.
6. Stir with the thermometer and observe the temperature change at regular intervals as it solidifies.

Safety

Wear eye protection.

Questions

1. When a liquid turns into a solid is the process exothermic or endothermic?
2. When all the liquid has turned into solid the temperature begins to drop. Why is this?

Classic chemistry experiments

RS•C

6. Accumulator

Topic

Reversible reactions, electricity from chemicals, energy in chemistry.

Timing

45 min.

Description

Students investigate the charging and discharging of a lead acid cell.

Apparatus and Equipment (per group)

- ▼ 100 cm^3 Beaker
- ▼ Power supply DC (4.5 V)
- ▼ Two wires
- ▼ Four crocodile clips
- ▼ Bulb in holder (1.25 V).

Chemicals (per group)

- ▼ Sulfuric acid 0.5 mol dm^{-3} (**Irritant**)
- ▼ Thin sheets of lead (2 cm x 8 cm) (degreased by soaking in sodium hydroxide solution (0.5 mol dm^{-3}) (**Corrosive**) for 10 minutes).

Teaching tips

If equations for the reactions are given some explanation will be required. Students can then identify which direction electrons travel during charging and discharging.

The experiment can be extended using data logging sensors to provide information on the way that the accumulator discharges. Connect a current sensor in series with the bulb (or connect a voltage sensor across its terminals) and monitor the changes on a computer screen. This can be used as an end-of-lesson demonstration to provide students with a good 'feel' of the discharge of the cell. The shape of the graph, which is a clue as to where a lead-acid cell is useful, is a point for discussion.

Background theory

As cell is charged:

$Pb(s) + 2H_2O(l) \rightarrow PbO_2(s) + 4H^+(aq) + 4e^-$ (anode)

$4H^+(aq) + 4e^- \rightarrow 2H_2(g)$ (cathode)

As cell is discharged:

$PbO_2(s) + 4H^+(aq) + 2e^- \rightarrow Pb^{2+}(aq) + 2H_2O(l)$ (anode)

$Pb(s) \rightarrow Pb^{2+}(aq) + 2e^-$ (cathode)

Safety

Wear eye protection. Wash hands after handling lead.

Accumulator

Introduction

Some types of cell are rechargeable. These cells store electricity. The most common rechargeable cell is the lead-acid type, which is the basis of car batteries. This experiment illustrates the charging and discharging of a lead-acid cell.

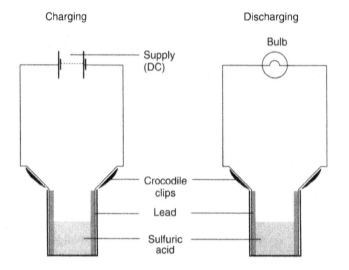

What to record

Complete the table:

Charging time /s	Time bulb is lit /s
180	
210	
240	
270	
300	

What to do

1. Connect the apparatus as shown.
2. Charge the cell at 4.5 V for three minutes.
3. Connect the cell for discharge.
4. Time how long the cell keeps the bulb lit.
5. Recharge the cell for a longer time and see how long the bulb stays lit.
6. Wash hands after handling lead.

Safety

Wear eye protection. Care with sulfuric acid.

Questions

1. Draw a line graph of your results. Charging time along the horizontal (x) axis and time lit along the vertical (y) axis.

7. Electricity from chemicals

Topic

Electricity and chemistry.

Timing

45 min.

Description

Students record the electromotive force produced when various pairs of metals are placed in sodium chloride solution.

Apparatus and equipment (per group)

- ▼ 100 cm^3 Beaker
- ▼ Galvanometer or voltmeter (0–3 V)
- ▼ Two wires
- ▼ Two crocodile clips.

Chemicals (per group)

- ▼ Sodium chloride solution
- ▼ Access to strips or rods of various metals:
    - Zinc
    - Copper
    - Iron
    - Lead
    - Magnesium.

Teaching tips

Data logging sensors and software can be used in this experiment to provide a large screen display of the voltage changes. Connect a voltage sensor across the electrodes and get the software to show the reading using a meter or graph.

Background theory

Metals high in the reactivity series have a tendency to release electrons to form ions. Metals low in the series do not readily form ions, and their ions easily form metal atoms. With zinc and copper:

$$Zn(s) \rightarrow Zn^{2+}(aq) + 2e^-$$

$$Cu^{2+}(aq) + 2e^- \rightarrow Cu(s)$$

Safety

Wear eye protection. Wash hands after handling lead.

Answers

1. Magnesium, zinc, iron, lead, copper.

Electricity from chemicals

Introduction

Reactive metals form ions more readily than less reactive metals. This experiment illustrates the tendency of various metals to form ions. Two different metals and an electrolyte form a cell. The more reactive metal becomes the negative pole from which electrons flow.

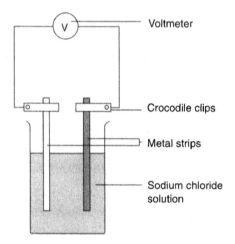

What to record

Complete the table.

What to do

1. Set up the apparatus as shown.
2. Record the voltage.
3. Try all the combinations of metals.
4. Wash hands after handling lead.

Safety

Wear eye protection.

Metals used	Which metal forms the positive terminal (+ve)	Which metal forms the negative terminal (-ve)	Voltage (V)
Zinc and copper			
Copper and lead			
Lead and iron			
Zinc and lead			
Iron and magnesium			
Zinc and iron			
Zinc and magnesium			
Lead and magnesium			
Copper and magnesium			
Copper and iron			

Questions

1. Place zinc, magnesium, copper, lead, and iron in order of reactivity.

RS•C

8. Iron in breakfast cereal

Topic

Separation, magnetism, chemical analysis.

Timing

30 min.

Description

Students examine cereal with a magnet. They extract the iron from a sample of the cereal.

Apparatus and equipment (per group)

- ▼ Mortar and pestle
- ▼ 250 cm^3 Beaker
- ▼ Very strong magnet (neodymium magnet). These magnets are available as electronics parts and are also recoverable from old computer disk drives.

Chemicals (per group)

- ▼ Iron-fortified breakfast cereal (eg Kellogg's® cornflakes).

Teaching tips

One suggestion is to begin by asking students to try to make the cereal stick to the magnet or move the flakes round the table. The cereal does not stick and friction is too great to allow the flakes to move on the table or bench surface. Elicit from students the idea to float the cereal, then hand out the worksheet. Instruct the students to keep the magnet away from the powdered cereal.

Background theory

Iron reacts with acid in the stomach and is eventually absorbed through the small intestine. If all the iron from the body were extracted, there would be enough for two small nails. Iron is essential for the production of haemoglobin.

Answers

1. No.
2. Fatigue, reduces resistance to disease, and increases heart and respiratory rates.

Iron in breakfast cereal

Introduction

Many breakfast cereals are fortified with iron. This iron is metallic and is added to the cereal as tiny particles of food grade iron before packaging. This experiment involves extracting the iron.

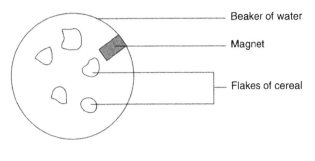

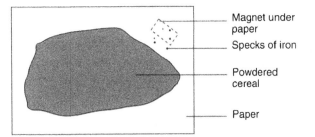

What to do

1. Float four to six pieces of cereal on the surface of a beaker of water.
2. Hold a magnet close to the cereal and see if this can cause a piece to move.
3. Put some cereal into a mortar and use a pestle to produce a very fine powder.
4. Spread the powder on a piece of paper.
5. Put a magnet under the paper and move the paper over the magnet.
6. Observe closely in the region of the magnet as the cereal moves over it.

Questions

1. Are all metals attracted to a magnet?
2. What are the symptoms of iron deficiency in the diet?

9. Unsaturation in fats and oils

Topic

Organic chemistry, saturated and unsaturated fats.

Timing

45 min.

Description

The students titrate different oils and fats mixed with Volasil against bromine water.

Apparatus and equipment (per group)

- White tile
- Conical flask
- Dropper pipette. (Use the type of teat pipette usually fitted to Universal Indicator bottles, that do not allow squirting – *eg* Griffin.)
- Burette (filled with bromine water)
- Boss
- Clamp
- Stand.

Chemicals (per group)

- Bromine water (**Harmful and irritant**) 0.02 mol dm^{-3} (This concentration does not have to be accurate, but if the concentration is reduced, less fat will be required to ensure sensible volumes of bromine water are used.)
- Volasil (Volasil 244 from BDH) (This is a mixture of organic chemicals which can act as a solvent for this experiment.)
- Cooking oil (animal)
- Cooking oil (vegetable)
- Olive oil.

Teaching tips

Other fats can be tried – *eg* melted butter, melted lard, melted margarine, and specific products such as Flora and Clover.

This experiment has also been trialled using $KMnO_4$(aq) (0.0005 mol dm^{-3}) as the indicator. This turns from purple to colourless while unsaturation is still present. The procedure is the same as for bromine water, but portions of the potassium permanganate are added with swirling until the mixture fails to produce a colourless solution. The mixture requires more and more swirling as the amount of potassium permanganate increases. Warming fats in the Volasil using a beaker of hot water helps the fat dissolve and also speeds up the reaction.

This experiment should be done in a fume-cupboard with ready filled burettes.

Background theory

Saturation and unsaturation.

Safety

Wear eye protection.

Answers

1. Depends on what is supplied.
2. Weighing the fats and oils and calculating the exact amount of bromine water used per mole.
3. Unsaturated compounds contain double covalent bonds.

Unsaturation in fats and oils

Introduction

Advertisements often refer to unsaturated fats and oils. This experiment gives a comparison of unsaturation in various oils.

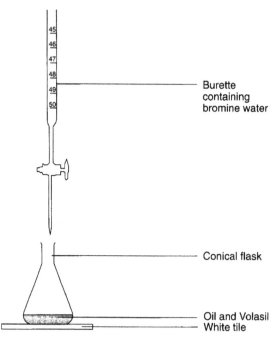

What to record

Volume of bromine water required for each oil.

What to do

1. Using a teat pipette, add five drops of olive oil to 5 cm^3 of Volasil in a conical flask.
2. Use a burette filled with a dilute solution of bromine water (0.02 mol dm^{-3}) **(Harmful and irritant)**. Read the burette.
3. Run the bromine water slowly into the oil solution. Shake vigorously after each addition. The yellow colour of bromine disappears as bromine reacts with the oil. Continue adding bromine water to produce a permanent yellow colour.
4. Read the burette. Subtract to find the volume of bromine water needed in the titration.
5. Repeat the experiment with: five drops of cooking oil (vegetable) and five drops of cooking oil (animal).

Safety

Wear eye protection.

Questions

1. Which sample is the most saturated and which is the most unsaturated?
2. This comparison is only approximate. How could the method be improved?
3. What does unsaturated mean?

10. The pH scale

Topic

Acids and alkalis.

Timing

30–45 min.

Description

Students measure the pH of various substances using Universal Indicator.

Apparatus and equipment (per group)

- Test-tubes
- pH chart
- 100 cm^3 Beaker
- Spatula.

Chemicals (per group)

- Limewater (calcium hydroxide solution) — 0.02 mol dm^{-3}
- Soda water (carbonic acid solution)
- Vinegar (ethanoic acid solution) — 0.05 mol dm^{-3}
- Milk of magnesia (magnesium oxide)
- Lemon juice (citric acid solution)
- Bicarbonate of soda (sodium hydrogencarbonate)
- Deionised water
- Lime (calcium hydroxide)
- Dilute sulfuric acid solution — 0.5 mol dm^{-3} or less
- Ammonia solution — 0.5 mol dm^{-3}
- Salt (sodium chloride)
- Universal Indicator bottle with integral pipette.

Teaching tips

It is advisable to demonstrate how to tell the colour of the indicator in coloured solutions by holding up to the light or dropping the indicator on the top of the solution and observing the colour at the interface.

Safety

Wear eye protection.

Background theory

Indicators.

Answers

1. Acids – *eg* soda water, vinegar, sulfuric acid, lemon juice; alkalis – *eg* milk of magnesia, lime, limewater, ammonia, bicarbonate of soda; neutral: salt.
2. Unlike Universal Indicator, litmus solution does not measure the pH.
3. Neutralisation should occur giving a green colour (**NB** exact colour/pH depends on concentrations).

RS•C

The pH scale

Introduction

The pH of a substance can be found by dissolving a small amount of the substance in deionised water and adding a few drops of Universal Indicator solution. The colour produced is compared with a pH chart.

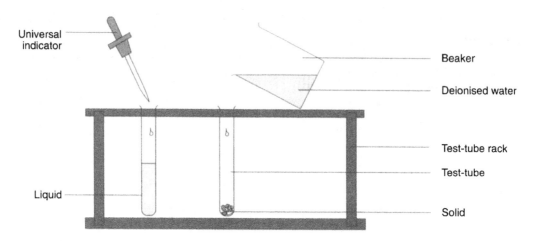

What to record

Prepare a table for your results

Solution	Colour with Universal Indicator	pH

What to do

1. Place one spatula measure of solid, or pour a few drops of liquid into a test-tube.
2. Half-fill the test-tube with deionised water from a small beaker, and shake to dissolve the solid or mix the liquid.
3. Add a few drops of Universal Indicator to the test-tube. Make a note of the colour in the table. Compare it against the pH colour chart and record the pH of the nearest colour in the table.

Safety

Wear eye protection.

Questions

1. List the substances that were acidic, substances that were alkaline and substances that were neutral.
2. Why might a scientist prefer to use Universal Indicator rather than a different indicator like litmus?
3. What would happen if equal amounts of vinegar and limewater were mixed?

11. The preparation and properties of oxygen

Topic

Reduction, transition metals, oxygen.

Timing

30 min.

Description

Students produce oxygen by heating potassium manganate(VII).

Apparatus and equipment (per group)

- Test-tube holder
- Ceramic wool
- Test-tube
- Spatula
- Bunsen burner
- Splints
- Heat-proof mat.

Chemicals (per group)

Potassium manganate(VII) (**Oxidising and harmful**)

Teaching tips

As an extension, students could add half a spatula measure of manganese(IV) oxide (**Harmful**) to 20 cm^3 (20 volume) hydrogen peroxide (**Irritant**) solution at room temperature and test for oxygen.

Safety

Wear eye protection. When $KMnO_4$ is heated, tiny particles shoot out. These are trapped by the ceramic wool.

Answers

1. $KMnO_4$.

The preparation and properties of oxygen

Introduction

Potassium manganate(VII) produces oxygen when heated. In this experiment oxygen is produced and identified with a glowing splint.

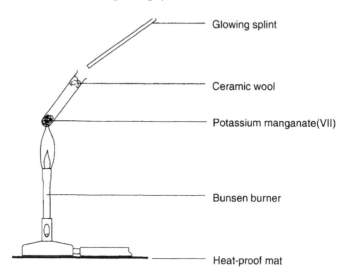

What to record

What was done and what was observed.

What to do

1. Place two spatula measures of potassium manganate(VII) in a test-tube.
2. Place a small piece of ceramic wool near the top of the test-tube. This stops fine dust escaping.
3. Gently heat the test-tube containing the potassium manganate(VII).
4. Light a splint and extinguish it, to make a 'glowing splint'.
5. Place the glowing splint just above the top of the test-tube. Keep heating the test-tube. The splint should relight.
6. Scrape out the ceramic wool. Let the test-tube cool to room temperature and then wash it out.
7. Notice the colours produced when the test tube is washed out.

Safety

Wear eye protection.

Potassium manganate(VII) is harmful if swallowed. It assists fire.

Questions

1. What is the chemical formula for potassium manganate(VII)?

12. Identifying polymers

Topic

Polymers.

Timing

45 min.

Description

The students place samples of plastic into solutions of known density to identify the polymers.

Apparatus and equipment (per group)

- ▼ Six test-tubes
- ▼ Test-tube rack
- ▼ Glass rod
- ▼ Samples of seven different plastics.

Chemicals (per group)

- ▼ Solutions 1–6 (see below for preparation)

Samples of:

Polymer	Density range/g cm^{-3}
EPS – expanded polystyrene	0.02–0.06
PP – polypropylene	0.89–0.91
LDPE – low density polyethylene	0.91–0.93
HDPE – high density polyethylene	0.94–0.96
PS – polystyrene	1.04–1.11
PVC – polyvinyl chloride	1.20–1.55
PET – polyethylene terephthalate	1.38–1.40

NB The names commonly used in industry are given here.

Sources of polymers

The following product types can be made from the named polymers:

HDPE – plastic bottles for milk, fruit juices, household cleaners and chemicals. Motor oil containers, some carrier bags and most aerosol caps.

LDPE – Jif lemon juice container. Some squeezy containers for sauces, cosmetics and plastic films – *ie* shrink wrap, sacks, freezer bags, carrier bags that are not crinkly, disposable pipettes, some aerosol caps, some plant pots and ink-tubes in ball-point pens.

PVC – plastic bottles for mineral water, fruit squash, cooking oil and shampoo. Sandwich and cake packs, food packaging trays, DIY blister packs, baby care product containers, cling film, ring-binder covers, records and watch straps.

PS – yoghurt pots, margarine tubs, clear egg boxes, food packaging trays, plastic cutlery and cups, clear plastic glasses, ball-point pen cases, cassette boxes and plastic coathangers

EPS – fast food packaging, meat packaging trays and egg boxes.

PP – plastic straws, containers for soft cheeses and fats, some margarine tubs, microwaveable food tubs and trays, film bags for crisps, biscuits and snacks, ketchup bottles and bottle caps.

PET – most plastic bottles for fizzy drinks, ovenproof food trays and roasting bags, audio and videotape.

For different coloured suggestions:

Container	Colour	Polymer
Low fat soft cheese tub	Blue	Polypropylene
Lemonade bottle	Colourless and clear	Polyethylene terephthalate
Fabric conditioner bottle	Pink	High density polyethylene
Shampoo bottle	Green	Polyvinyl chloride
Yoghurt pot	White	Polystyrene
Jif lemon juice container	Yellow	Low density polyethylene
Burger box	Gold	Expanded polystyrene

Solution	Density	Composition of 1000 cm^3 solution
1	0.79	Pure ethanol (IMS)
2	0.91	471 g (596 cm^3) ethanol in 439 cm^3 deionised water
3	0.94	354 g (448 cm^3) ethanol in 586 cm^3 deionised water
4	1.00	Deionised water
5	1.15	184 g K_2CO_3 in 965 cm^3 deionised water
6	1.38	513 g K_2CO_3 in 866 cm^3 deionised water

A class of 30 students requires approximately 90 cut samples of each polymer. Containers should therefore be easily and quickly cut. Samples should be about 4 mm square and labelled with the name of the container, not the polymer. If all the containers are not different colours, samples could be cut into different shapes. Samples can be recovered at the end of the lesson.

Teaching tips

Do not give out the questions (with results) until students have completed the experiment.

Background theory

Polymer name	Acronym	Colour	Fingerprint 1	2	3	4	5	6
Polyethylene terephthalate	PET	any	S	S	S	S	S	S
Polyvinyl chloride	PVC	any	S	S	S	S	S	–
Polystyrene	PS	any	S	S	S	S	–	–
High density polyethylene	HDPE	any	S	S	S	–	–	–
Low density polyethylene	LDPE	any	S	S	–	–	–	–
Polypropylene	PP	any	S	–	–	–	–	–
Expanded polystyrene	EPS	any	–	–	–	–	–	–

Samples sink if their density is greater than the density of the solution into which they are put. Students can collect waste plastics over several weeks before this lesson. Product containers and their lids are not always made out of the same material so it is important to check their identities prior to the lesson. The identity of polystyrene is confirmed if samples float in a solution of density 1.15 g cm^{-3} but sink in a solution of density 1.00 g cm^{-3}. The solutions will need to be made in advance so the plastics can be tested in advance of the lesson. Solutions should be labelled with their densities (given here in g cm^{-3} at 20 °C). Temperature changes affect densities of solutions so they should be stored together and not in a cold store. Instructions for preparing solutions are as follows for 1 dm^3 of each. A class of 30 students will use approximately 200 cm^3 of each.

Safety

Wear eye protection.

Ethanol is highly flammable and toxic. Potassium carbonate solid is an irritant, as are concentrated solutions.

Answers

1. To dislodge air bubbles, which may make a 'sinker', float.
2. Depends on what is provided.

Identifying polymers

Introduction

In this experiment solutions with known densities are used to identify the polymers used in everyday materials.

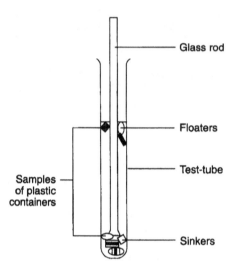

What to record

Sample	Colour (or shape)	Solutions					
		1	2	3	4	5	6

What to do

1. Fill six test-tubes with solutions 1 to 6 and label each tube.
2. Place a sample of each type of polymer into solution 1.
3. Use a glass rod to stir the contents of the tube. Observe whether the waste plastics float or sink.
4. For samples that sink, write the letter S in column 1 of the results table.
5. Wash the glass rod and dry it on a tissue or paper towel.
6. Repeat the test for solutions 2 to 6. Use a new sample each time.

Safety

Wear eye protection.

Solutions 1,2 and 3 are highly flammable and toxic. Solutions 5 and 6 are irritants.

Questions

1. Why were the solutions stirred once the plastics were added?
2. Use the following table to identify the plastics. Fill in the table.

Polymer name	Acronym	Colour	Fingerprint					
			1	2	3	4	5	6
Polyethylene terephthalate	PET	Any	S	S	S	S	S	S
Polyvinyl chloride	PVC	Any	S	S	S	S	S	–
Polystyrene	PS	Any	S	S	S	S	–	–
High density polyethylene	HDPE	Any	S	S	S	-	–	–
Low density polyethylene	LDPE	Any	S	S	–	-	–	–
Polypropylene	PP	Any	S	–	–	–	–	–
Expanded polystyrene	EPS	Any	–	–	–	–	–	–

Polymer name	Acronym	Sample
Polyethylene terephthalate	PET	
Polyvinyl chloride	PVC	
Polystyrene	PS	
High density polyethylene	HDPE	
Low density polyethylene	LDPE	
Polypropylene	PP	
Expanded polystyrene	EPS	

RS•C

13. Energy values of food

Topic

Energy in food.

Timing

45–60 min.

Description

Students burn various foods of known mass. They heat a known volume of water and calculate the amount of energy in the food.

Apparatus and equipment (per group)

- ▼ Stirring thermometer
- ▼ Boss, clamp and stand
- ▼ Test-tube/metal calorimeter
- ▼ Access to balance
- ▼ Bunsen burner
- ▼ Mounting needle
- ▼ Teaspoon.

Chemicals (per group)

- ▼ Different foods.

Teaching tips

Mini-marshmallows, crisps, pasta, bread, potatoes, bacon, broad beans (dried) and cheese can be used.

Students could weigh any unburnt food.

Data logging sensors and software can be used in this experiment to provide a large screen display of the temperature change. If the food stops burning too soon and the tube of water starts to cool, the graph will show a brief drop in temperature. This potential source of error makes a good discussion point – in fact the temperature drop can be accounted for in the calculation.

Safety

Wear eye protection

Do not permit tasting of foods.

Answers

1. Some heat is lost to the surroundings, not all the food may have burnt.
2. Exothermic.
3. 16380 J. Heat losses to surroundings will produce a lower figure from experiment.

Energy values of food

Introduction

In this experiment various foods are tested to find how much energy they contain.

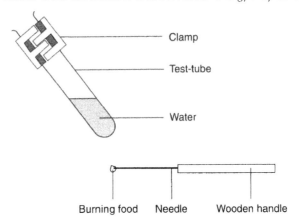

What to record

Measurement	Food		
Mass/g			
Temperature of water before heating/°C			
Temperature of water after heating/°C			
Change in temperature/°C			
Heat absorbed by water/J (Temperature change x 4.2)			
Heat absorbed by water per gram of food/J			

What to do

1. Put 10 cm³ of water in a test-tube. Clamp the test-tube in the retort stand at an angle as shown in the diagram.
2. Weigh a small piece of food and record the mass in your table.
3. Take the temperature of the water in the test-tube and record it in the table.
4. Fix the food on the end of the mounted needle. If the food is likely to melt when heated put it on a teaspoon instead of on the needle.
5. Light the food using a bunsen burner. As soon as the food is alight, hold it about 1 cm below the test-tube. If the flame goes out, quickly relight it.
6. When the food stops burning, stir the water in the test-tube with the thermometer and note the temperature. Record it in your table.
7. Empty the test-tube and refill it with another 10 cm³ of water. Repeat the experiment using a different food each time.

Safety

Wear eye protection.

Questions

1. Suggest reasons why this experiment may not be a fair test?
2. Burning gives out heat. What is the name given to this sort of reaction?
3. The label on a packet of cheese says 100 g provides 1638 kJ. Calculate how many joules this is per gram of cheese and compare it to the cheese in your experiment. (1 kJ = 1000 J)

14. A compound from two elements

Topic

Mixtures and compounds.

Timing

30–45 min.

Description

Students note properties of a mixture of iron and sulfur and then heat them to form iron sulfide. The properties of the new compound are then compared with the mixture.

Apparatus and equipment (per group)

- ▼ Tongs
- ▼ Bunsen burner
- ▼ Heatproof mat
- ▼ Magnet.

Chemicals (per group)

- ▼ Sulfur in a sealed plastic bag
- ▼ Iron filings in a sealed plastic bag
- ▼ Mixture of iron filings and sulfur in a sealed plastic bag
- ▼ Ignition tube not more than a quarter full containing a mixture of iron and sulfur. (Seven parts iron to four parts sulfur by mass)

Teaching tips

Instruct students not to allow the magnet near the ignition tubes before heating otherwise iron filings get stuck to the magnet.

Background theory

Heating iron with sulfur forms strong chemical bonds, that hold the atoms tightly together.

Safety

Wear eye protection. Consider the use of a fume cupboard. If a fume cupboard is not used, good ventilation is essential. Asthmatics should use a fume cupboard.

Answers

1. Iron + sulfur → iron sulfide.
2. Chemical bonds are formed to hold iron and sulfur together.
3. FeS.

A compound from two elements

Introduction

A mixture of iron and sulfur can easily be separated. This is because there are no chemical bonds between the sulfur and the iron. The iron is magnetic and is therefore easily removed from the sulfur. In this experiment, a mixture of iron and sulfur are heated to make a new compound.

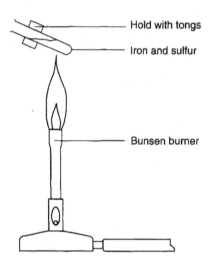

What to do

1. Examine the plastic bag of sulfur, the bag of iron and the bag containing a mixture of the two.
2. Run a magnet over each of the bags.
3. Set up the apparatus as shown in the diagram.
4. Light a Bunsen burner and half open the air-hole to give a medium flame.
5. Heat the very end of the tube strongly. When the mixture starts to glow, move the Bunsen burner to one side.
6. Watch the mixture in the tube. (If the glow just goes out, heat the tube again.)
7. Let the tube cool down completely.
8. The substance from the tube is a new compound called iron sulfide.
9. Test the iron sulfide with a magnet. Does the magnet pick it up?

Safety

Wear eye protection. Do not get too close to the fumes.

Questions

1. Write a word equation for this reaction.
2. What has happened to the iron and the sulfur in forming iron sulfide?
3. What is the chemical formula for iron sulfide?

15. Chemistry and electricity

Topic

Acids and alkalis, electrolysis of salt solution.

Timing

30–45 min.

Description

Students make up a salt solution with indicator and complete an electrical circuit. The cations / anions are attracted to the carbon electrode causing the indicator to change colour.

Apparatus and equipment (per group)

- ▼ Plastic petri dish
- ▼ Filter papers
- ▼ 6 V battery or power pack
- ▼ Leads and crocodile clips
- ▼ Carbon electrode
- ▼ Dropping pipette. Use the type of teat pipette (usually fitted to Universal Indicator bottles) that does not allow squirting – eg Griffin.

Chemicals (per group)

- ▼ Sodium chloride
- ▼ Universal Indicator
- ▼ Methyl orange.

Teaching tips

Other indicators to try might include: bromocresol green (lead attached to positive terminal), screened methyl orange (try both terminals), blue litmus (positive) and red litmus (negative). Phenolphthalein does not work very well in this experiment.

Background theory

When the 'pencil' is attached to the negative lead, H^+ ions are attracted to it, producing the colour associated with acids for that particular indicator. If the 'pencil' is attached to the positive lead, the reverse happens.

Safety

Wear eye protection.

Answers

1. When attached to the negative lead the writing is red, when attached to the positive lead it is purple.
2. H^+ ions are attracted to the negative electrode, OH^- ions are attracted to the positive electrode. So depending on which electrode the pencil is attached to it will affect the colour of the indicator and therefore the writing.

RS•C

Chemistry and electricity

Introduction

In this experiment, electricity and some indicators are used to make coloured writing.

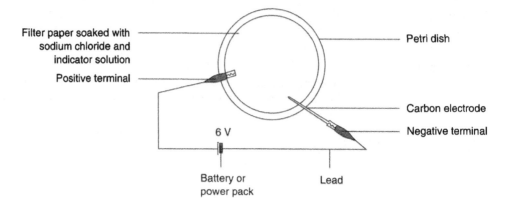

What to do

1. Dissolve a spatula measure of sodium chloride in 2 cm^3 of water. Add three drops of methyl orange indicator.
2. Lay a filter paper inside a plastic petri dish. Drop the solution onto the paper using a dropping pipette, until the paper holds no more solution.
3. Attach the positive end of a 6 V battery to a lead ending in a crocodile clip. Use the crocodile clip to grip one end of the paper.
4. Attach the negative end of the battery to a carbon electrode.
5. Write lightly on the wet paper, using the carbon electrode. What colour is the writing?
6. Repeat the experiment using Universal Indicator. Describe what happens.

Safety

Wear eye protection.

Questions

1. What would happen if the lead were attached to the positive electrode using Universal Indicator? Try this if there is time.
2. Explain what reactions have occurred to produce the colours.

16. Combustion

Topic

Combustion.

Timing

30 min.

Description

A candle is burnt inside a gas jar. The students then test for the presence of carbon dioxide and water.

Apparatus and equipment (per group)

- ▼ Gas jar and lid
- ▼ Candle on a tray
- ▼ Heat-proof mat.

Chemicals (per group)

- ▼ Limewater 0.02 mol dm^{-3}
- ▼ Blue cobalt chloride paper.

Teaching tips

As an extension, the students could suggest other experiments to do to find out if other fuels form carbon dioxide and water when they burn.

Data logging sensors and software can be used to demonstrate what may be happening in the jar as the candle burns. Use a bell jar and place sensors inside to monitor humidity, temperature, light or oxygen levels as the candle burns. The software will show the changes as a graph against time. When the candle extinguishes, readmit air to the jar and continue to record for a few moments.

Safety

Wear eye protection. Some fuels might be quite hazardous. A thorough risk assessment should be done.

Answers

1. Oxygen.
2. Carbon dioxide and water.
3. Methane or similar hydrocarbon or fuel.

Combustion

Introduction

Hydrocarbons produce carbon dioxide and water when they burn. In this experiment the products of combustion are captured and tested.

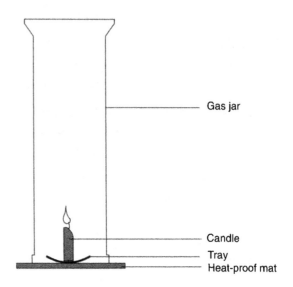

What to record

What was done and what was observed.

What to do

1. Set up the apparatus as shown in the diagram. The gas jar should be placed over the lit candle on a heatproof mat.
2. When the candle goes out, put a lid on the gas jar.
3. Test to see if the candle made water by adding a piece of blue cobalt chloride paper, test the sides of the jar. If it turns pink, water is present.
4. Now test to see if carbon dioxide was produced. Pour a little limewater into the gas jar. Swill it around a little. If carbon dioxide is present, the limewater turns cloudy.

Safety

Wear eye protection.

Questions

1. What is the gas that reacts with the hydrocarbon when it burns?
2. What gases does the candle produce when it burns?
3. Name another fuel that produces the same gases when it burns.

17. The determination of relative atomic mass

Topic

Moles, relative atomic mass, stoichiometry.

Timing

45 min.

Description

Students react magnesium with dilute hydrochloric acid and collect the hydrogen produced in an inverted burette.

Apparatus and equipment (per group)

- ▼ Burette
- ▼ Burette stand
- ▼ 250 cm^3 Beaker.

Chemicals (per group)

- ▼ Hydrochloric acid 2 mol dm^{-3} (**Irritant**)
- ▼ Magnesium ribbon (0.02–0.04 g)(~3.5 mm standard ribbon)(**Flammable**).

Teaching tips

It is advisable to demonstrate this method beforehand. The inversion is not difficult. Rest the end of the burette on the lip of the beaker and swing the tap end round and upward to a vertical position. A finger over the end is not necessary. It is important that the liquid level starts on the graduated scale of the burette. If the liquid level is not on the scale a momentary opening of the tap allows the liquid to drop onto the scale. Do not fold the magnesium ribbon; bend the ribbon and push it into the burette ensuring it remains under tension.

A further suggestion is to use a spreadsheet to pool the class results.

Background theory

The volume of one mole of gas at stp (0 °C, 101 500 N m^{-2}) is 22.4 dm^3.

(At room temperature and average pressure an answer of about 24 dm^3 can be expected.)

Students that are more able may be able to use the equation

$P_1V_1/T_1 = P_2V_2/T_2$ to find volume at stp. The temperature and pressure in the laboratory should be measured.

If the mass of magnesium used is 0.03 g then the volume of hydrogen collected would be ~30 cm^3.

Safety

Wear eye protection. Teachers should control how the magnesium ribbon is distributed.

The determination of relative atomic mass

Introduction

One mole of any gas occupies the same volume when measured under the same conditions of temperature and pressure. In this experiment, the number of moles of hydrogen produced from a known mass of magnesium is measured. The relative atomic mass of magnesium can therefore be calculated.

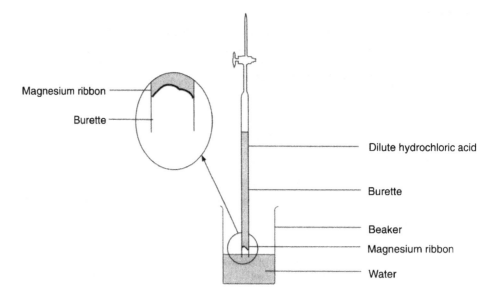

What to record

The mass of magnesium used and the volume of hydrogen produced.

What to do

1. Clean a piece of magnesium ribbon (about 3.5 cm long) and weigh it accurately (This should weigh between 0.02 g and 0.04 g; if not adjust the amount used.)
2. Measure 25 cm^3 of dilute hydrochloric acid into the burette. Carefully add 25 cm^3 of water on top of this.
3. Push the magnesium in the end of the burette so it stays in position with its own tension.
4. Add 50 cm^3 of water to a 250 cm^3 beaker.
5. Quickly invert the burette into the water, (if this is done quickly and carefully very little will be lost. It is important that the liquid level in the burette starts on the graduated scale. If it is not on the scale; momentarily open the tap, this allows the level to drop).
6. Clamp the burette vertically.
7. Take a burette reading (**NB:** it is upside down!)
8. Observe how the magnesium reacts as the acid diffuses downwards, wait until all the magnesium has reacted.
9. Note the new volume on the burette (**NB:** it is upside down).
10. Record your results.

Safety

Wear eye protection.

Questions

The equation for the reaction is

Mg + 2HCl → MgCl$_2$ + H$_2$

1. Copy out and fill in the gaps:

    ____ g magnesium was produced from ____ cm^3 of hydrogen.

    ____ g magnesium was produced from 1 cm^3 of hydrogen

    ____ g magnesium was produced from 24000 cm^3 of hydrogen.

    ____ g magnesium would be produced from 1 mole of hydrogen.

This is the mass of 1 mole of magnesium. Numerically, this number is the relative atomic mass of magnesium.

18. The reaction of a Group 7 element (iodine with zinc)

Topic

Halogens, transition metals, salt formation, direct synthesis.

Timing

45 min.

Description

React zinc with iodine at room temperature to form a salt.

Apparatus and equipment (per group)

- Spatula
- Stirring thermometer
- Two test-tubes
- Test-tube rack
- Tripod
- 250 cm^3 Beaker.

Chemicals (per group)

- Iodine crystals (**Harmful**)
- Zinc powder (about a grain of rice amount) (**Flammable**)
- Ethanol (**Highly flammable**)
- Boiling water (access to electric kettle or thermostatically controlled water bath).

Teaching tips

The iodine solution should become 'clear' when the reaction is complete. Provide zinc and iodine crystals for each group in separate little pots so students can tip these straight into the alcohol. Alternatively the iodine could already be placed in the test-tubes, to avoid any possible skin contact. Iodine could even be provided in solution.

This is a versatile reaction. With some adaptation, this reaction could also be used to determine empirical formulae, demonstrate an electrical cell and to study heat energy from chemical energy (enthalpy change).

Background theory

Reactions of Group 7 elements. Salt formation.

Safety

Wear eye protection. Ensure the room is well ventilated and free from sources of ignition.

Answers

1. Halogen.
2. Salt producer.
3. Zinc + iodine → zinc iodide.

The reaction of a Group 7 element (iodine with zinc)

Introduction

This experiment involves producing a salt by reacting a Group 7 element (iodine) with zinc. This is an example of salt preparation by direct synthesis.

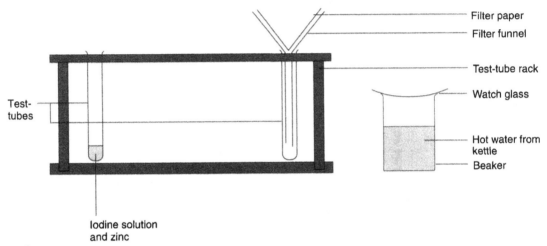

What to do

1. Use the measuring cylinder to add 4 cm^3 of alcohol to a test-tube.
2. Dissolve a few crystals of iodine in the alcohol.
3. Note the temperature.
4. Add some powdered zinc (about the volume of a grain of rice) to the iodine solution.
5. Stir with the thermometer and note the highest temperature.
6. Record any changes which suggest a chemical reaction has taken place.
7. Filter the liquid into another test-tube (the excess zinc is left behind).
8. Collect a beaker of hot water, and place a watch glass on the top
9. Pour the liquid filtrate onto the watch glass and allow the alcohol to evaporate. Observe the product.
10. The product is classified as a SALT.

Safety

Wear eye protection. Do not touch the iodine.

Questions

1. What is the collective name for the Group 7 elements?
2. What does this name mean?
3. Complete the word equation:
   Zinc + iodine →

19. Reactions of halogens

Topic

Halogens.

Timing

30 min.

Description

Students use halogen solutions to see what happens with Universal Indicator paper. Students then carry out displacement reactions.

Apparatus and equipment (per group)

- ▼ White tile
- ▼ Universal Indicator paper
- ▼ Test-tubes
- ▼ Test-tube rack.

Chemicals (per group)

- ▼ Potassium bromide solution — 0.2 mol dm^{-3}
- ▼ Potassium iodide solution — 0.2 mol dm^{-3}
- ▼ Potassium chloride — 0.2 mol dm^{-3}
- ▼ Chlorine solution 0.02 mol dm^{-3} (**Irritant solution, liberates toxic gas**), Bromine solution 0.02 mol dm^{-3} (**Harmful and irritant**) Iodine solution 0.02 mol dm^{-3}.
  The accuracy of the strength of these dilute solutions is not important for the reactions. A saturated aqueous solution of chlorine is approximately 0.09 mol dm^{-3}.

Teaching tips

Sodium halides could be used instead of potassium.

Background theory

Chlorine reacts with water to give hydrochloric acid and bleach (chloric (I) acid). Chlorine solution bleaches all the indicator paper, bromine bleaches just the spot where it was placed, iodine faintly bleaches the spot. Halogens get less reactive as the group is descended. Chlorine displaces bromine from potassium bromide and iodine from potassium iodide. Bromine also displaces iodine from potassium iodide, no reaction occurs with iodine and potassium bromide or potassium chloride.

Safety

Wear eye protection. Care with chlorine, bromine and iodine solutions. Do not allow asthmatics to inhale chlorine gas.

Answers

1. Chlorine solution is the strongest bleaching agent.
2. Chlorine is also the most reactive, then bromine, then iodine.
3. Chlorine + potassium bromide → potassium chloride + bromine
   $$Cl_2 + 2KBr \rightarrow 2KCl + Br_2$$

Reactions of halogens

Introduction

The Group 7 elements are called the halogens. This experiment involves some reactions of the halogens.

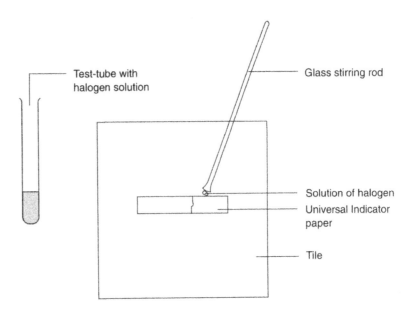

What to record

Complete the table.

	Effect on indicator paper	Reaction with potassium chloride solution	Reaction with potassium bromide solution	Reaction with potassium iodide solution
Chlorine water				
Bromine water				
Iodine water				

What to do

1. Put a piece of Universal Indicator paper onto a white tile.
2. Use a glass stirring rod to transfer a few drops of the first solution onto the indicator paper.
3. Repeat this with a fresh piece of paper and the different solutions.
4. In a test-tube, add some chlorine solution to a solution of potassium bromide.
5. Add some chlorine solution to a solution of potassium iodide.
6. Now try mixing solutions of bromine and potassium iodide. If there is time, mix the other combinations of solutions to complete the table.

Safety

Wear eye protection. Do not breathe chlorine gas.

Questions

1. Which halogen solution is the strongest bleaching agent?
2. Which halogen is the most reactive?
3. Write a word and symbol equation for the reaction of chlorine with potassium bromide.

20. The sublimation of air freshener

Topic

Sublimation, separation, purification.

Timing

30 min.

Description

Students use an air freshener in a water bath to demonstrate sublimation.

Apparatus and equipment (per group)

- ▼ Two 100 cm^3 beakers
- ▼ Shallow dish
- ▼ Thermometer
- ▼ Access to a fume cupboard.

Chemicals (per group)

- ▼ Small piece of solid air freshener
- ▼ Ice
- ▼ Hot water.

Teaching tips

It is possible to use other materials that sublime. Solid toilet bowl cleaners work best, but if cheap ones that contain *p*-dichlorobenzene are used, handle them with tongs in the fume cupboard as this substance is harmful. If iodine is used, use only a few crystals and do the activity in a fume cupboard. Iodine is also harmful. Naphthalene (**Harmful**) mothballs must be heated to near 70 °C to sublime. Dry ice sublimes at –78.5°C and above. If possible use a coloured air freshener, notice that the material that collects on the cold beaker is white. The dye does not sublime because it is not chemically a part of the compound that does sublime. Vapour deposition is an important industrial process for separation and purification.

Gel type air fresheners – *eg* Glade – do not work.

Background theory

Sublimation is the vaporisation of a solid. The opposite process, the formation of a solid directly from a vapour, is called deposition. The heat from the water bath causes the solid air freshener to sublime. The cold beaker causes the vaporised air freshener to condense and re-form the solid.

Safety

Wear eye protection. Most of these substances are harmful. In day to day use at room temperature, this does not present a problem, as the vapour pressure is relatively low. However, if heated until they sublime, vapour levels could reach hazardous concentrations. Hence, either a fume cupboard should be used or some other method of preventing escape into the air.

Answers

1. The air freshener does not sublime measurably below this temperature.
2. Sublimation is the physical change that occurs when a substance goes from a solid phase directly to a gaseous phase.
3. Particles go from being close-packed and ordered in a solid to separated and disordered in a gas.

Classic chemistry experiments

RS•C

The sublimation of air freshener

Introduction

Sublimation is an interesting physical change. When a substance sublimes, it changes directly from a solid to a gas without passing through the liquid state. Dry ice sublimes, as do iodine and mothballs. This experiment involves the study of another common substance that sublimes – air freshener.

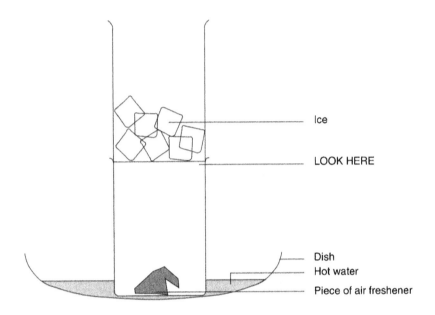

What to do

1. Place a few lumps of air freshener in the bottom of the 100 cm^3 beaker.
2. Put the 100 cm^3 beaker carefully on top of the other 100 cm^3 beaker.
3. Fill the top beaker three quarters full with ice. Ensure no ice enters the beaker below.
4. Fill the shallow dish or pan about one-third full of hot water (at a higher temperature than 45 °C).
5. Place the sublimation apparatus in the shallow dish in a fume cupboard.
6. Observe what happens to the solid. Be patient, it may take a while.

Safety

Wear eye protection. Use a fume cupboard.

Questions

1. What might be the significance of 45 °C? Try lower and higher temperatures if there is time.
2. Define 'sublimation'.
3. Use the particle theory of matter to explain what is happening and include a particle diagram.

21. Testing the pH of oxides

Topic

Metal and non-metal oxides.

Timing

20–30 min.

Description

Students test the pH of various metal and non-metal oxides.

Apparatus and equipment (per group)

▼ Test-tubes.

Chemicals (per group)

Provide solutions of :

▼ Nitric acid (labelled Nitrogen oxide and water)   0.2 mol dm^{-3} (**Irritant**)
▼ Sodium hydroxide (labelled Sodium oxide and water)   0.2 mol dm^{-3} (**Irritant**)
▼ Potassium hydroxide (labelled Potassium oxide and water) 0.2 mol dm^{-3} (**Irritant**)
▼ Phosphoric acid (labelled Phosphorus (V) oxide and water) 0.2 mol dm^{-3}
▼ Calcium hydroxide (labelled Calcium oxide and water)   0.2 mol dm^{-3}
▼ Universal Indicator (in bottle with dropping pipette).

In the past, chemistry teachers have labelled these as oxide solutions. It is understood that these are not solutions, the oxides do not dissolve – they have reactions with water. While it is important to maintain accuracy, it is suggested that an '…and water' approach is taken. This is accurate and it allows the teacher to focus on the teaching point without the need to cover another set of reactions. However, teachers may wish to cover these reactions with water, this may depend on the age and ability of the class.

Teaching tips

Students are often confused with the difference between bases and alkalis.

Background theory

In general metal oxides are basic, non-metal oxides are acidic. Some metal oxides react with water to form alkaline solutions. Some metal oxides do not react with water. They test neutral in water because they are insoluble but are bases and react with acids. Non-metal oxides react with water to form acids.

Safety

Wear eye protection. If other concentrations of solution are used, ensure they are less than 0.5 mol dm^{-3}.

Answers

1. Sodium, potassium and calcium oxides are metallic.
2. All the metal oxides are basic.
3. Some metal oxides do not react to form hydroxides in water therefore they are neutral.

Testing the pH of oxides

Introduction

In this experiment the pH of various oxides is tested.

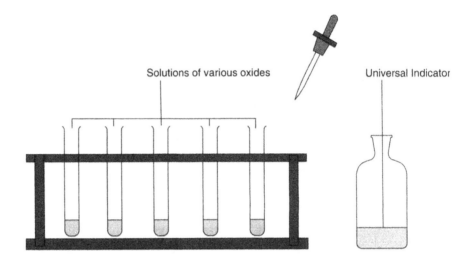

What to record

Name of oxide	Colour of Universal Indicator	pH value	Acid, alkali or neutral
Nitrogen oxide			
Sodium oxide			
Potassium oxide			
Phosphorus(V) oxide			
Calcium oxide			

What to do

1. Using separate test-tubes, collect a sample (about 2 cm^3) of each oxide in water.
2. Add three drops of Universal Indicator solution to each sample.
3. Record the results in a table showing the oxide, the colour of the Universal Indicator, the pH and whether the oxide is acidic, alkaline or neutral in water.

Safety

Wear eye protection

Questions

1. Which compounds in the table are metal oxides?
2. Comment on your results for the metal oxides?
3. Some metal oxides do not react with water. Predict the pH of these compounds.

RS•C

22. Exothermic or endothermic?

Topic

Energy transfer.

Timing

30 min.

Description

Students look at four reactions to see if they are exothermic or endothermic.

Apparatus and equipment (per group)

- ▼ Beakers
- ▼ Thermometer.

Chemicals (per group)

- ▼ 10 cm^3 Sodium hydroxide solution   0.4 mol dm^{-3} (**Irritant**)
- ▼ 10 cm^3 Dilute hydrochloric acid   0.4 mol dm^{-3} (**Irritant**)
- ▼ 10 cm^3 Sodium hydrogen carbonate solution   0.4 mol dm^{-3} (**Irritant**)
- ▼ Four spatula measures citric acid
- ▼ 10 cm^3 Copper(II) sulfate solution   0.4 mol dm^{-3} (**Irritant**)
- ▼ Four spatula measures magnesium powder (**Highly flammable**)
- ▼ 3 cm Magnesium ribbon (**Highly flammable**)
- ▼ 10 cm^3 Dilute sulfuric acid   0.4 mol dm^{-3} (**Irritant**).

Teaching tips

Students get confused about endothermic reactions. An endothermic reaction drops in temperature as it takes in (or absorbs) heat. A temperature sensor attached to a computer can be used in place of a thermometer in this experiment. The software can plot the temperature change on a graph and for a demonstration, show it on a screen-size digital display.

Background theory

Bond making and bond breaking.

Safety

Wear eye protection.

Answers

1. Neutralisation.
2. Carbon dioxide.
3. Displacement reaction – magnesium is higher in the reactivity series so it displaces copper from its sulfate.
4. Sodium hydrogen carbonate solution + citric acid is endothermic, the other reactions are exothermic.
5. Reactions involve making and breaking bonds. Energy is required to break bonds. Energy is released when bonds are made. If more energy is required to break the bonds in a reaction than that released by making the bonds then the reaction is endothermic (heat is absorbed). If less energy is required to break the bonds than that released by making the bonds then the reaction is exothermic (heat is produced).

Exothermic or endothermic?

Introduction

Some reactions give out heat and others take in heat. In exothermic reactions the temperature goes up, in endothermic reactions the temperature goes down. In this experiment, various reactions are examined. Temperatures are measured to decide whether a particular reaction is exothermic or endothermic.

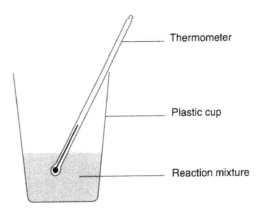

What to record

Complete the table

Reaction	Temperature before mixing/°C	Temperature after mixing/°C	Exothermic or endothermic
Sodium hydroxide solution + dilute hydrochloric acid			
Sodium hydrogen carbonate solution + citric acid			
copper(II) sulfate solution + magnesium powder			
Dilute sulfuric acid + magnesium ribbon			

What to do

1. Use the apparatus as shown.
2. Put 10 cm^3 of sodium hydroxide solution in the beaker, record the temperature then add 10 cm^3 of dilute hydrochloric acid, stirring with the thermometer. Record the maximum or minimum temperature.
3. Repeat the procedure for the following reactions: (a) sodium hydrogen carbonate solution and citric acid; (b) copper(II) sulfate solution and magnesium powder; and (c) dilute sulfuric acid and magnesium ribbon.

Safety

Wear eye protection. Some of the solutions are irritant.

Questions

1. The first reaction is between an acid and an alkali, what do we call this type of reaction?
2. Which gas is given off when sodium hydrogen carbonate reacts with citric acid?
3. Which type of reaction takes place between copper(II) sulfate and magnesium?
4. Which reactions are exothermic and which are endothermic?
5. Describe in terms of bond breaking and bond making, why some reactions are exothermic and some are endothermic.

23. Water expands when it freezes

Topic

Change of state.

Timing

10–15 min in lesson. Overnight freeze. 10–15 min to examine result.

Description

This basic experiment shows students that ice occupies a larger volume than its equivalent as water. This expansion is used to explain various events such as burst pipes and the freeze-thaw process in rocks. This is when water finds its way into cracks. When this water freezes it expands and can split off a piece of rock.

Apparatus and equipment (per group)

- ▼ Glass bottle with screw cap
- ▼ Plastic bag
- ▼ Tie.

Teaching tips

Care is needed with broken glass. This works well with thin glass bottles.

Background theory

Explain freeze-thaw weathering (see Description above).

Safety

Care with broken glass.

Answers

1. Water in pipes can expand in cold weather, causing them to burst.
2. The milk can freeze overnight and shatter the bottle or push the top off.
3. See description above.

RS•C

Water expands when it freezes

Introduction

Water expands when it freezes. Most liquids contract when they freeze so this property of water is unusual. This property is clearly shown in this experiment. This process is used to explain how ice can break rocks apart.

What to record

What happens.

What to do

1. Fill a screw top bottle right up to the top with water.
2. Screw on the top tightly.
3. Tie the sealed bottle in a clear plastic bag.
4. Leave overnight in a freezer.

Safety

Care when removing frozen bottle from freezer.

Questions

1. Use what happened to explain why water pipes sometimes burst in winter
2. What happens when your milk freezes on the doorstep in winter?
3. Use this knowledge to add captions to the diagram that explain how rocks crack.

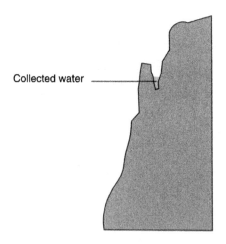
Collected water

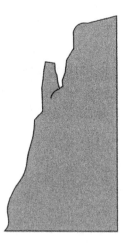

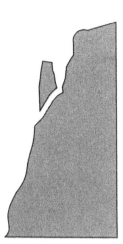

24. The chemical properties of the transition metals – the copper envelope

Topic

The transition metals.

Timing

30 min.

Description

Students fold copper foil, heat it and observe any reaction. Students then open up the folded metal and look at the areas of copper which were inaccessible to the air.

Apparatus and equipment (per group)

- ▼ Bunsen burner
- ▼ Heat resistant mat
- ▼ Tongs.

Chemicals (per group)

- ▼ Copper foil (4 cm square).

Teaching tips

Copper, like many transition metals, only reacts slowly with oxygen in the air.

When heated copper forms a layer of black copper oxide on its surface.

$$2Cu(s) + O_2(g) \rightarrow 2CuO(s)$$
Copper   Oxygen   Copper oxide

Background theory

Oxidation.

Safety

Wear eye protection. Care – sharp corners on copper. Care – copper stays hot for some time.

Answers

1. No.
2. Black surface.
3. Oxygen.
4. Copper oxide.
5. The inside is still shiny, the inside was inaccessible to oxygen so no oxidation occurred.

The chemical properties of the transition metals – the copper envelope

Introduction

Transition metals are situated between Groups 2 and 3 of the Periodic Table. They have important uses. One well-known transition metal is copper. Transition metals have similar reactions and properties.

Sc	Ti	V	Cr	Mn	Fe	Co	Ni	Cu	Zn
Y	Zr	Nb	Mo	Tc	Ru	Rh	Pd	Ag	Cd
La	Hf	Ta	W	Re	Os	Ir	Pt	Au	Hg

This is the sequence of folds to produce the copper envelope.

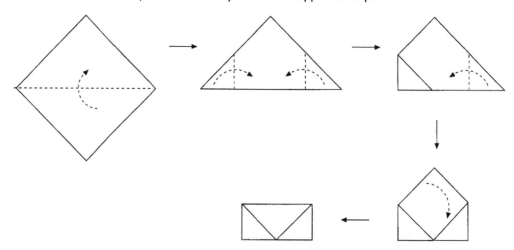

What to do

1. Fold a square of copper foil into an envelope as shown in the diagram. Care – sharp corners
2. Hold the envelope in tongs.
3. Heat strongly in a Bunsen burner flame for 5 min.
4. Place the copper envelope on the heatproof mat to cool. Care – hot.
5. When the envelope is cool enough to hold; open it up and compare the inside with the outside.

Safety

Wear eye protection.

Questions

1. Does the copper burst into flames like magnesium?
2. What does the copper look like after it has cooled?
3. Which gas in the air has copper reacted with?
4. What is the black coating on the surface called?
5. What is the appearance of the inside of the envelope and why is this?

RS•C

25. The reactivity of Group 2 metals

Topic

Periodicity, reactivity, salt formation, reactions of acids.

Timing

30 min.

Description

Students react magnesium and calcium with hydrochloric acid to find out which is the most reactive.

Apparatus and equipment (per group)

- ▼ Test-tube rack
- ▼ Two test-tubes
- ▼ Splint.

Chemicals (per group)

- ▼ Hydrochloric acid 1 mol dm^{-3} (**Irritant**)
- ▼ Magnesium (small piece of ribbon) (**Highly flammable**)
- ▼ Calcium small piece (**Flammable**) (Do not use old calcium, use fresh stock.)

Teaching tips

Discussion about how to judge the speed of the reaction is advisable. Remind students about the test for hydrogen. Calcium can be distributed on pieces of filter paper.

Background theory

Group 1 is the most reactive group of metals. The Group 1 metals get more reactive the lower they are in the group. Group 2 metals are also reactive. This experiment compares their reactivity.

Safety

Wear eye protection.

Answers

1. Calcium.
2. Magnesium + hydrochloric acid → magnesium chloride + hydrogen
   Calcium + hydrochloric acid → calcium chloride + hydrogen
3. $Mg + 2HCl \rightarrow MgCl_2 + H_2$
   $Ca + 2HCl \rightarrow CaCl_2 + H_2$

The reactivity of Group 2 metals

Introduction

Metals in Group 2 of the Periodic Table are less reactive than those in Group 1. This experiment indicates the relative reactivity of elements within the group.

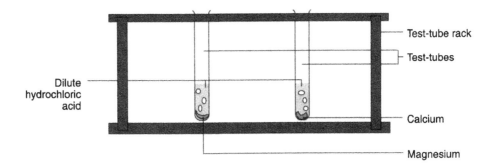

What to do

1. Fill two test-tubes a quarter full with dilute hydrochloric acid.
2. Into one test-tube drop a small piece of magnesium
3. Into the other, drop a small piece of calcium.
4. Compare the reactivity of the two metals.
5. Drop another bit of magnesium into the first test-tube and put your thumb over the end.
6. When the pressure can be felt, take your thumb off and test the gas with a lighted splint.
7. Record what happens.

Safety

Wear eye protection.

Questions

1. Which is the more reactive, magnesium or calcium?
2. Write word equations for these reactions.
3. Write formula equations for these reactions.

RS•C

26. Melting and freezing

Topic

Change of state, solids, liquids and gases, physical changes.

Timing

45–60 min.

Description

Students heat stearic acid and measure the temperature, observing when it melts, then allow it to cool, record the temperature and note when it starts to solidify. They then plot a graph.

Apparatus and equipment (per group)

- ▼ Test-tube
- ▼ Graph paper
- ▼ 250 cm^3 Beaker
- ▼ Stop clock
- ▼ Tripod and gauze
- ▼ Spatula
- ▼ Thermometer (0–100 °C)
- ▼ Clamp, boss, and stand
- ▼ Bunsen burner.

Chemicals (per group)

- ▼ Stearic acid (octadecanoic acid).

Teaching tips

Energy must be supplied to melt a solid, this same energy is released when the liquid resolidifies. Remind students not to attempt to move the thermometer in the solid stearic acid, as it will break. Collect in the test-tubes with thermometers frozen and these can be released later by melting.

This presents a good opportunity to demonstrate how to maintain a steady temperature using a Bunsen burner. This can be achieved by sliding the Bunsen burner aside as the boiling becomes too vigorous, slide it back as the water stops boiling. It is not essential that the water bath is boiling. Students can be provided with another thermometer, and asked to maintain a lower temperature, say 80 °C.

Napthalene is used for this experiment because stearic acid does not give a sharp melting or freezing point as it is rarely pure. Naphthalene is much better, but steps must be taken to prevent the escape of naphthalene vapour. If naphthalene is used, the experiment should be done in a fume cupboard and a ceramic wool plug inserted in top of the boiling tube.

A temperature sensor attached to a computer can be used in place of a thermometer. It can plot the temperature change on a graph and show this as it occurs. A slight modification of the experiment can yield an intriguing result: when the test-tube is cooling place it in an insulated cup containing a few cm^3 of water. Use a second temperature sensor to monitor the temperature of the water. The water temperature should rise as the stearic acid cools and it should continue to rise even as it changes state.

Background theory

States of matter, particulate theory of matter.

Safety

Wear eye protection.

Answers

1. About 60 °C
2. The same as melting temperature.
3. The temperature stays the same for a while, both solid and liquid are present, this is the melting/freezing point. When the solid is melting the energy is used to separate the particles from each other. When the liquid is freezing, energy is given out as forces begin to hold the particles together.

RS•C

Melting and freezing

Introduction

In this experiment, a solid turns into a liquid and then the liquid turns into a solid. The energy changes are examined.

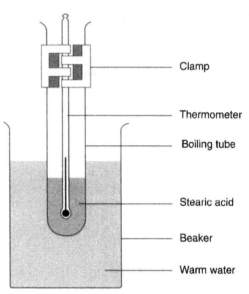

What to record

Complete the table.

Time/min	Temperature/°C

What to do

1. Fill a 250 cm³ beaker with about 150 cm³ tap water.
2. Heat it on a tripod and gauze until the water just starts to boil.
3. Set up the apparatus as shown in the diagram and start the timer.
4. Try and maintain the temperature of the water. It should be just boiling but not boiling vigorously.
5. Record the temperature every minute as the stearic acid heats up, until it reaches about 70 °C. Show in your table the temperature where the solid starts to melt.
6. Use the clamp stand to lift the tube from the hot water. Record the temperature every minute as the stearic acid cools down until it reaches about 50 °C. Note the temperature in your table when the first signs of solid formation are observed.
7. Plot a line graph of your results. Put time along the bottom and temperature up the side. Label your graph to show where stearic acid is a solid, a liquid or present in both states.

Safety

Wear eye protection.

Questions

1. What is the melting point of stearic acid?
2. What is the freezing point of stearic acid?
3. Why are there flat sections on your graph? Explain this in terms of the forces between particles.

RS•C

27. Diffusion in liquids

Topic

Diffusion.

Timing

30 min.

Description

Students react lead nitrate and potassium iodide. They add a crystal of each to a petri dish of deionised water.

Apparatus and equipment (per group)

- ▼ Petri dish
- ▼ Forceps.

Chemicals (per group)

- ▼ Deionised water
- ▼ Crystal of lead nitrate (**Toxic**)
- ▼ Crystal of potassium iodide.

Teaching tips

It is advisable to inform students of what to look for. Students may recognise the particular shade of yellow from the yellow lines used on the side of roads in the UK.

Background theory

Lead nitrate and potassium iodide solutions react to give 'clouds' of solid yellow lead iodide.

Safety

Wear eye protection. Lead nitrate is toxic. Use forceps to handle crystals.

Answers

1. Lead nitrate + potassium iodide → potassium nitrate + lead iodide
2. $PbNO_3 + KI \rightarrow KNO_3 + PbI$
3. Particles of the dissolved substance can move between the particles of solvent.
4. The iodide ion moves fastest. The lead particles are bigger (and have more mass) than the iodide ions. Since they are at the same temperature, they have the same energy, therefore the iodide ions must be moving faster.

Diffusion in liquids

Introduction

Diffusion occurs in liquids but more slowly than in gases. The particles are not as free to move about. This experiment illustrates diffusion in a liquid.

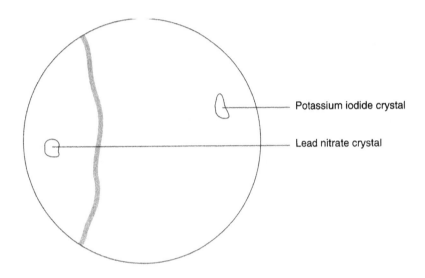

What to do

1. Fill a petri dish with deionised water.
2. Use forceps to drop crystals of lead nitrate and potassium iodide on opposite sides of the dish.
3. The solid crystals form solutions that react. Observe what happens.
4. Watch diffusion occurring, as a yellow solid slowly forms between the two crystals.

Safety

Wear eye protection.

Questions

1. Write a word equation for the reaction.
2. Write a formula equation for the reaction.
3. How can this diffusion process be explained?
4. With crystals of lead nitrate and potassium iodide in water, the yellow solid forms closer to the lead nitrate crystal. Which reactant diffuses faster, the lead ion or the iodide ion? Why?

RS•C

28. Chemical filtration

Topic

Adsorption, separation.

Timing

45 min.

Description

Students observe how specially treated carbon removes colour and odour from various solutions.

Apparatus and equipment (per group)

- ▼ Funnel
- ▼ Test-tube
- ▼ Filter paper
- ▼ Test-tube rack.

Chemicals (per group)

- ▼ Potassium manganate(VII)
- ▼ Decolourising carbon (activated charcoal)
- ▼ Ink or food colouring
- ▼ Sauerkraut, dill pickle juice, or vinegar.

Teaching tips

1. Adsorption by charcoal is also used to remove unburned hydrocarbons from car exhausts, harmful gases from the air, and unwanted colours from certain products.
2. Students may find the difference between adsorption and absorption confusing. Adsorption: a gas, liquid, or a dissolved substance is gathered on the surface of another substance – eg charcoal. Absorption: a liquid is soaked up, as with a blotter. It is taken in completely and mixes with the absorbing material – eg absorbent cotton.
3. Carbon reduces the purple MnO_4^- (Mn(VII)) ion in manganate(VII) to the colourless Mn^{2+} (Mn(II)) ion.
4. Charcoal powder is very black and very messy.
5. Juice from sauerkraut, dill pickle or vinegar still smells after filtration, but noticeably less.

Background theory

Heating wood to a very high temperature in the absence of air makes charcoal. When it is heated to an even higher temperature, about 930 °C, impurities are driven from its surface and it becomes activated charcoal, sometimes called decolourising charcoal. This activated charcoal can remove impurities in either the gaseous or liquid state from many solutions. It does so by the process of adsorption, or by attracting these molecules to the surface.

Safety

Wear eye protection.

Answers

1. In each case the filtrate is colourless and clear and does not have the impurities in the original solution.
2. Charcoal has many small holes. This feature gives it a large surface area. This large surface area allows it to attract a large number of molecules of the impure substances.
3. Water is filtered through charcoal to remove impurities that would otherwise discolour and give a bad taste to drinking water.

Chemical filtration

Introduction

In this experiment, carbon that has undergone special treatment to make it into decolourising carbon is shown to remove colour and odour from various solutions. This form of carbon is sometimes called activated charcoal. This method is used to remove objectionable taste and odours from drinking water.

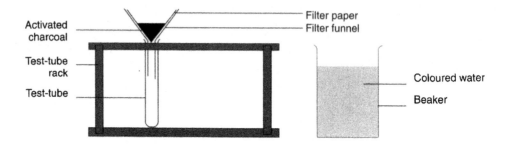

What to do

1. Fold a piece of filter paper, place it in a funnel, and put the stem of the funnel into a test-tube in a test-tube rack.
2. Add about five spatulas of decolourising carbon to the funnel
3. Add one drop of ink or food colour to 100 cm^3 of water in a beaker.
4. Carefully pour some of the coloured water onto the charcoal in the filter paper.
5. Prepare another filter paper with the same amount of carbon. This time filter a solution made by adding two or three crystals of potassium manganate(VII) to 100 cm^3 of water.
6. Repeat the activity, this time filter sauerkraut juice, dill pickle juice or vinegar.

Safety

Wear eye protection.

Questions

1. Describe the material before and after filtration in each of the three activities.
2. How does carbon remove colour and odour?
3. How could this process be used to provide pure water for drinking?

29. Rate of reaction – the effects of concentration and temperature

Topic

Kinetics.

Timing

30 min.

Description

Students react potassium iodate and a starch solution. They vary the concentration and temperature to affect the reaction time.

Apparatus and equipment (per group)

- Two 250 cm^3 beakers
- Water bath (or some means of warming solution A).

Chemicals (per group)

- Solution A – 4.3 g of KIO_3 per dm^3 (**Oxidising solid**)
- Solution B – starch solution

Make the starch solution as follows: Make a paste of 4 g of soluble starch in a small amount of warm water. Slowly add 800 cm^3 of boiling water. Boil for a few minutes then cool the solution. Add 0.2 g of sodium metabisulfite ($Na_2S_2O_5$) (**Harmful solid**). Add 5 cm^3 of 1.0 mol dm^{-3} sulfuric acid (**Irritant**). Dilute to 1 dm^3.

Teaching tips

The colour change takes about 5–6 minutes. A colorimeter sensor or a light sensor set up as a colorimeter can be used to monitor colour change on the computer. The result, in the form of graphs on the computer, provides very useful material for analysis using data logging software. While a colorimeter sensor is ideal, it is easy to substitute a light sensor clamped against a plastic cuvette filled with the reactants. The data logging software should clearly show the change occurring on a graph. Measure the rate of change by using its slope or the time taken for a change to occur.

Background theory

The mechanism is not clearly understood but the following simplified sequence has been proposed.

1. IO_3^- reacts with HSO_3^- to form I^-:
   $IO_3^- + 3HSO_3^- \rightarrow I^- + 3H^+ + 3SO_4^{2-}$
2. I^- reacts with IO_3^- to form I_2.
3. I_2 is immediately consumed by reacting with HSO_3^-:
   $I_2 + HSO^{3-} + H_2O \rightarrow 2I^- + SO_4^{2-} + 3H^+$
4. When all of the HSO^{3-} has been used up, I_2 accumulates.
5. Iodine reacts with starch to form a coloured complex.

Safety

Wear eye protection.

Answers

1. There are more molecules of reactant in the solution therefore more chance of reacting.

2. Increasing the temperature has two effects. Since the particles are moving faster, they will travel a greater distance in a given time and so will be involved in more collisions. In addition, because the particles are moving faster a larger proportion of the collisions will exceed the activation energy, the energy needed to react. The rate of the reaction therefore increases.

3. Depending on the results of the experiment, increase/decrease concentration to a specific amount AND/OR increase/decrease the temperature by a specific amount.

Rate of reaction – the effect of concentration and temperature

Introduction

In this experiment, two colourless solutions are mixed to make a solution which becomes dark blue. Changing the concentration or temperature of the solutions changes the time required for the blue colour to develop.

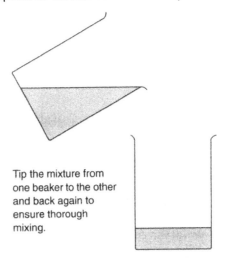

Tip the mixture from one beaker to the other and back again to ensure thorough mixing.

What to record

The conditions and the times for reactions to occur.

What to do

1. Place 50 cm^3 of solution A in a 250 cm^3 beaker.
2. Place the same volume of solution B in a second beaker.
3. Mix the two solutions by pouring from one beaker into the other several times.
4. Note the time required for a reaction to occur (formation of blue colour).
5. Repeat, but use solution A that has been diluted to one half the concentration. Note the time for the reaction to occur.
6. Repeat using solution A warmed to 35 °C. Note the time for a reaction to occur.

Safety

Wear eye protection

Questions

1. Why does increasing the concentration usually result in an increased rate of reaction?
2. Why does increasing the temperature usually result in an increased rate of reaction?
3. How could this experiment be set up so it took exactly 10 min to turn blue?

30. Reaction between carbon dioxide and water

Topic

Indicators, acids and bases, gas liquid reactions, atmospheric chemistry, acid rain.

Timing

30 min.

Description

Students will blow or speak into a conical flask containing one of two solutions and observe any changes.

Apparatus and equipment (per group)

▼ Two 250 cm^3 conical flasks.

Chemicals (per group)

▼ Ethanol (95 per cent) (**Highly flammable**)

Access to:

▼ Thymolphthalein indicator with dropper
▼ Phenol red indicator with dropper
▼ Sodium hydroxide solution with dropper 0.4 mol dm^{-3} (**Irritant**)

Use the kind of bottles normally used for indicators, the type with dropping pipettes in the lid that do not allow squirting – eg Griffin

Teaching tips

Swirling speeds up the reaction. Straws are not required, simply breathing or speaking into the flask will cause the indicator to change colour.

Background theory

Eventually, CO_2 from the students' breath will produce enough acid in the solution to change the colour of the indicator:

$$CO_2(g) + H_2O(l) \rightarrow H_2CO_3(aq) \rightleftharpoons H^+(aq) + HCO_3^-(aq)$$

CO_2 also reacts with NaOH. This reaction produces the less basic Na_2CO_3:

$$2NaOH(aq) + CO_2(g) \rightarrow Na_2CO_3(aq) + H_2O(l)$$

Safety

Wear eye protection.

Answers

1. There is not enough CO_2.
2. To ensure the solution is slightly alkali at the beginning and to absorb any initial CO_2, or any other acid.

Reaction between carbon dioxide and water

Introduction

When carbon dioxide reacts with water a weak acid is formed. Carbon dioxide is present in exhaled breath. Observing a colour change using an acid-base indicator shows the reaction between carbon dioxide and water.

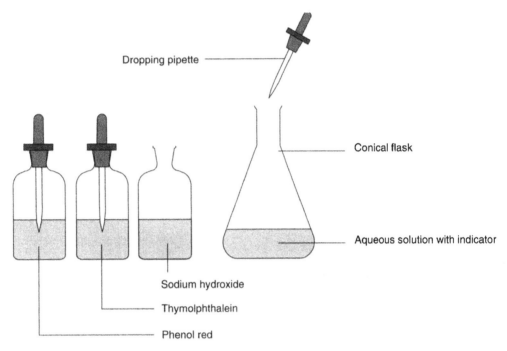

What to do

Activity 1
1. Place about 125 cm^3 of ethanol (**Highly flammable**) in a 250 cm^3 conical flask.
2. Add five or six drops of thymolphthalein indicator to the alcohol.
3. Add just enough dilute sodium hydroxide (**Irritant**) (dropwise) to produce a blue colour.
4. Talk or blow gently into the flask – ie add the carbon dioxide.
5. Continue adding the carbon dioxide until a colour change is observed.

Activity 2
1. Place about 125 cm^3 of water in a 250 cm^3 conical flask.
2. Add one or two drops of phenol red to the water.
3. Add two drops of sodium hydroxide solution (**Irritant**) to produce a red solution.
4. Talk or blow gently into the flask – ie add carbon dioxide.
5. Continue adding the carbon dioxide until a colour change is observed.

Safety

Wear eye protection.

Questions

1. Why does the colour change not occur instantly?
2. Why are a few drops of sodium hydroxide solution (NaOH) added before the experiment?

31. Competition for oxygen

Topic

Reactivity series.

Timing

60 min.

Description

Mixtures of metals and metal oxides are heated over a Bunsen burner flame. Students observe the reactions and decide if a reaction occurs.

Apparatus and equipment (per group)

- ▼ Bunsen burner
- ▼ Tripod
- ▼ Pipe clay triangle
- ▼ Crucible
- ▼ Tongs.

Chemicals (per group)

- ▼ Magnesium oxide and iron mixture
- ▼ Lead oxide and iron mixture
- ▼ Lead oxide and zinc mixture
- ▼ Copper(II) oxide and zinc mixture.

Teaching tips

This is an ideal opportunity to demonstrate the thermit reaction (see *Classic Chemistry Demonstrations*, p196. London: RSC, 1995). Some teachers recommend using ceramic paper rather than crucibles, as some products are difficult to remove from the crucible after reaction. One consideration is that using ceramic paper causes more mess and the Bunsen burners become clogged from spilt powders.

Some teachers like to heat the mixture directly from above. Care is needed not to let powder be sprayed by the flame.

Background theory

Students should understand the idea of a competition for oxygen. This can be used as an introduction to the reactivity series but it can also be used when students know how to predict the outcome of the reaction from the reactivity series.

Safety

Students must wear eye protection, and they must not lean over the reaction mixture.

Some of the reactions may be unexpectedly violent. Ensure the room is well ventilated or use a fume cupboard.

Answers

1.

Reaction mixture	Does this mixture react?
Magnesium oxide and iron	No
Lead oxide and iron	Yes
Lead oxide and zinc	Yes
Iron oxide and zinc	Yes

2. Lead oxide + iron → iron oxide + lead
   Lead oxide + zinc → zinc oxide + lead
   Iron oxide + zinc → zinc oxide + iron

Competition for oxygen

Introduction

This experiment involves the reaction of a metal with the oxide of another metal. When reactions like these occur, the two metals compete for the oxygen. The more reactive metal finishes up with the oxygen (as a metal oxide). If the more reactive metal starts as the oxide then no reaction takes place.

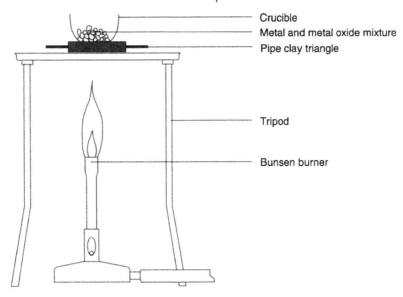

What to record

Decide whether a reaction takes place in each case.

What to do

1. Set up the apparatus as shown in the diagram.
2. Place one spatula measure of one of the reaction mixtures into the crucible.
3. Heat the mixture gently at first and then more strongly. Watch carefully to see what happens but do not lean over the crucible.
4. Allow the mixture to cool. Look for evidence that a reaction has taken place.
5. Use your observations to decide which of the two metals has 'won' the competition for oxygen – which is more reactive?
6. Choose another mixture and repeat the experiment.

Safety

Wear eye protection. Do not lean over the crucible.

Questions

1. Complete the following table.

Reaction mixture	Does this mixture react?
Magnesium oxide and iron	
Lead oxide and iron	
Lead oxide and zinc	
Iron oxide and zinc	

2. Write word equations for any reactions that occur.

32. Making a crystal garden

Topic

Silicates, earth science.

Timing

20–30 min (+ overnight for crystals to grow).

Description

Students prepare silicate crystals from warm sodium silicate solution and various metal nitrates or sulfates.

Apparatus and equipment (per group)

- ▼ Glass jar
- ▼ Watchglass
- ▼ Glass stirring rod
- ▼ Forceps.

Chemicals (per group)

- ▼ Hot deionised water
- ▼ Sodium silicate solution (**Irritant**)
- ▼ A few crystals of: cobalt nitrate (**Irritant**), nickel nitrate (**Harmful**), iron(III) nitrate (**Irritant**), manganese sulfate, magnesium nitrate.

Teaching tips

It should be emphasised to students that only a few small crystals are needed.

Background theory

Crystals. Silicon. Formation of silicates in the Earth's mantle.

Safety

Wear eye protection. Handle crystals with forceps.

Answers

1. The different colours are characteristic of the different metals present.
2. A high enough temperature could not be reached in the laboratory.

Making a crystal garden

Introduction

The formation of molten silicates in the Earth's mantle involves the formation of silicon dioxide and its subsequent reaction at high temperatures with metal oxides. In this experiment coloured silicates are formed in the laboratory.

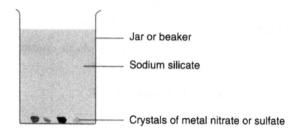

What to record

What is observed.

What to do

1. Pour sodium silicate solution (**Irritant**) into a glass jar to a depth of 3 cm. Add hot water to this solution, stirring well during the addition. The final depth of liquid required is about 12 cm. Stirring should continue until no separate silicate layer is visible.
2. Allow the solution to stand until the liquid is quite still.
3. Use forceps to drop a few crystals into the liquid, try and choose different colours. Ensure that the crystals do not fall close to each other.
4. Cover the jar and leave overnight.

Safety

Wear eye protection.

Questions

1. Why are the silicate crystals different colours?
2. Why could silicates not be formed in the laboratory in exactly the same way as they are formed in the Earth's mantle?

33. Extracting metal with charcoal

Topic

The extraction of metal oxides by reduction with carbon.

Timing

60 mins.

Description

Students react two metal oxides with carbon. They observe whether the metal is extracted from its oxide.

Apparatus and equipment (per group)

- Two ignition tubes
- Tongs
- Bunsen burner
- Heatproof mat
- Spatulas.

Chemicals (per group)

- Charcoal
- Lead(II) oxide (**Toxic** – may cause harm to unborn children and there is a danger of cumulative effects. Harmful if swallowed.)
- Copper(I) oxide (**Harmful**)

Teaching tips

Hold the ignition tubes with tongs. Copper(I) oxide gives better results than copper(II) oxide. Lead(II) oxide gives better results than lead(IV) oxide.

Background theory

Students should understand the basic concept of oxidation and reduction, and competition for oxygen. They should also understand the idea of an electrochemical or reactivity series.

Metal oxides are reduced by carbon to produce the pure metal. This is used in industry to separate metals from their ores. This method is used for those metals that are lower than carbon in the electrochemical series. The carbon is oxidised to carbon dioxide.

Safety

Wear eye protection. Ensure the room is well ventilated. Use only small amounts of oxides.

Answers

1. To allow the molten metal to solidify and prevent any reoxidation.
2. The carbon is oxidised to carbon dioxide.
3. $2PbO + 2C \rightarrow 2Pb + CO_2$
4. The metal is reduced and the carbon is oxidised.

Extracting metal with charcoal

Introduction

This extraction experiment consists of two competition reactions. A metal oxide is reacted with charcoal. If the charcoal (carbon) is more reactive it will remove the oxygen from the metal oxide and leave a trace of metal in the reaction vessel. Start with an oxide of lead, then observe what happens to an oxide of copper.

What to do

Lead oxide
1. Add one spatula of lead oxide (**Toxic**) to the ignition tube.
2. Add one spatula of charcoal powder.
3. Mix this mixture using an unfolded paper clip.
4. Strongly heat this mixture for five minutes in the Bunsen burner flame.
5. Allow to cool.
6. Tip the mixture onto a heatproof mat.

Copper oxide
1. Add one spatula of copper oxide (**Harmful**) to the ignition tube.
2. Carefully add one spatula of charcoal powder on top without any mixing.
3. Strongly heat these two layers for five minutes in the Bunsen burner flame.
4. Allow to cool and then look closely at the ignition tube where the powders meet.

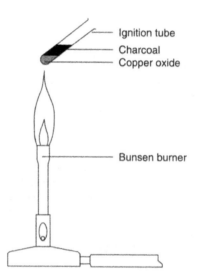

Safety

Wear eye protection.

Questions

1. Why should the mixture be cool before it is tipped out of the ignition tube at the end?
2. What happens to the carbon if it takes oxygen from the metal oxide?
3. Write the equation for the reaction of lead oxide.
4. Which element is oxidised and which is reduced in this reaction?

34. The migration of ions

Topic

Electrolysis.

Timing

45 min.

Description

Students perform two experiments. In the first they observe the migration of manganate(VII) ions towards the negative electrode. In the second experiment they observe two ions meeting as they migrate towards opposite electrodes. Silver chromate (an insoluble red compound) is formed.

Apparatus and equipment (per group)

- Two crocodile clips
- Microscope slides
- Two lengths of connecting wires
- Filter paper
- Source of DC (20 V)
- Forceps.

Chemicals (per group)

- Small crystal of potassium manganate(VII) (**Oxidising**)
- Silver nitrate solution 0.1 mol dm^{-3}
- Potassium chromate(VI) solution (5 per cent) (K_2CrO_4) (**Very toxic solid, toxic solution, may cause cancer**).

Teaching tips

This method should be demonstrated before the students carry out this experiment.

Teachers may choose to let their students carry out one or both experiments.

Background theory

The direct current is carried through the electrolyte by ions. At the electrodes, electrons are transferred to or from the ions. The processes that take place at the electrodes are thus either reduction or oxidation half-reactions. Electrolysis is thus a reduction-oxidation process.

Safety

Wear eye protection. Do not exceed 20 V.

Answers

1. Manganate(VII) ions are coloured.
2. Manganate(VII) ions are negative.
3. Silver ions migrate towards the negative electrode and chromate ions migrate towards the positive electrode.
4. $2Ag^+ + CrO_4^{2-} \rightarrow Ag_2CrO_4$

The migration of ions

Introduction

In an electrolysis experiment, the ions migrate towards electrodes of opposite charge. In the first part of this experiment the migration of manganate ions is observed.

In the second part of this experiment, silver and chromate ions meet as they migrate towards opposite electrodes. Silver chromate, an insoluble red compound is formed.

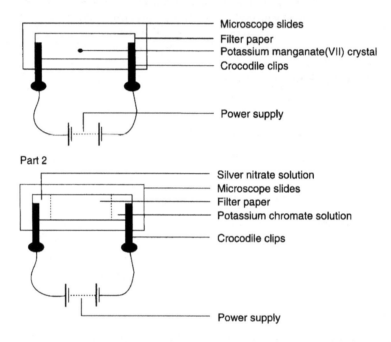

What to record

Draw a diagram to show the filter paper at the end of both experiments. Mark the ends of the paper with a + or − to show which terminal of the power supply each end is connected to.

What to do

Part 1
1. Cut a piece of filter paper slightly smaller than a microscope slide. Draw a faint pencil line across the middle.
2. Moisten the filter paper with tap water. Fasten the paper to the slide with crocodile clips.
3. Use forceps to put a small crystal of potassium manganate(VII) in the centre of the paper.
4. Connect the clips to a power supply set at not more than 20 V DC. Switch on and wait about ten minutes.

Part 2
1. Cut a piece of filter paper slightly smaller than a microscope slide. Draw a faint pencil line across the middle.
2. Moisten the paper with tap water. Fasten the paper to the slide with crocodile clips.

3. At the end of the paper where the positive electrode will be placed, moisten the paper with silver nitrate solution. (**Will stain fingers**)
4. At the end of the paper where the negative electrode will be placed, moisten the paper with potassium chromate solution. (**Toxic**).
5. Connect the clips to a power supply set at no more than 20 V DC. Switch on and wait about 10 min, or until you see a change.

Safety

Wear eye protection.

Questions

1. Potassium manganate (VII) consists of two ions - potassium ions and manganate (VII) ions. One of these ions is coloured. Which is it likely to be?
2. From the direction of movement, what does this indicate about the charge on the manganate ion?
3. What is happening in the second experiment?
4. The formula of silver nitrate is $AgNO_3$ and potassium chromate is K_2CrO_4. Write the ionic formula equation for the reaction.

RS•C

35. The reduction of iron oxide by carbon

Topic

Reactivity series, oxidation and reduction.

Timing

20–30 min.

Description

Students react iron(III) oxide with carbon on the end of a spent match to produce iron. The mixture is fused with sodium carbonate.

Apparatus and equipment (per group)

- ▼ Match
- ▼ Spatula
- ▼ Bunsen burner
- ▼ Magnet.

Chemicals (per group)

- ▼ Sodium carbonate (access to a small quantity to adhere to the end of a match)
- ▼ Iron(III) oxide (access to a small quantity to adhere to the end of a match).

Teaching tips

A fragment of metallic iron will be attracted to the magnet. Sodium carbonate fuses easily and brings the oxide into intimate contact with the carbon. One suggestion is that students test iron(III) oxide with a magnet before the reaction to show it is not magnetic.

Background theory

Reactivity series, reduction.

Safety

Wear eye protection.

Answers

1. Reduction can be described as the removal of oxygen.
2. Carbon comes between aluminium and iron in the reactivity series. The products of the reaction between carbon and zinc oxide.
3. Calcium is reactive and therefore its oxide is very stable, carbon is not reactive enough to displace oxygen from calcium oxide.

The reduction of iron oxide by carbon

Introduction

Metals high in the reactivity series will reduce the oxides of those lower in the series. The oxides of metals between zinc and copper in the reactivity series can be reduced by carbon. In this experiment, sodium carbonate is used to fuse the reactants in intimate contact.

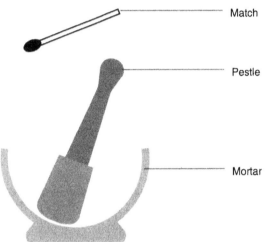

What to do

1. Char the point of a used match, moisten it with a drop of water and rub on some sodium carbonate crystals.
2. Rub the point in some powdered iron(III) oxide (Fe_2O_3) and heat in a blue Bunsen burner flame until the point glows strongly.
3. Allow to cool.
4. Crush the charred head in a mortar and pestle then run a magnet through the pieces.

Safety

Wear eye protection.

Questions

1. What does 'reduction' mean?
2. Carbon does not reduce aluminium oxide. Where would carbon be placed in this reactivity series?
   Potassium
   Sodium
   Calcium
   Magnesium
   Aluminium
   Zinc
   Iron
   Lead
   Copper

   What other information would you need to determine carbon's exact place.

3. Explain why calcium oxide cannot be reduced using carbon.

RS•C

36. Experiments with particles

Topic

Mixing of materials. Particle theory of matter.

Timing

45 min.

Description

Students measure any changes in volume when solids are mixed with solids, salt is dissolved in water, and alcohol is mixed with water.

Apparatus and equipment (per group)

- Stirring rod
- Two 100 cm^3 measuring cylinders
- Spatula.

Chemicals (per group)

- Sodium chloride
- Water
- Sand
- Dried peas
- Ethanol (**Highly flammable**).

Teaching tips

Ensure that the students understand what is meant by prediction.

Students should be encouraged to suggest explanations for their results, even if the inferences seem not to support their predictions
– *eg* 25 cm^3 peas + 25 cm^3 sand = 46 cm^3 total volume.

A student may feel that water and alcohol are continuous materials with the heavier one resting on the other – therefore squashing the lighter one. The effect would be to have a lower total volume.

Students may need reminding not to try and measure 25 cm^3 accurately but to pour approximately then take an accurate measurement.

Background theory

Particulate nature of matter.

Safety

Wear eye protection.

Answers

1. Both activities involve the mixing of particles, combined volumes are less than the sum of the parts.
2. Salt particles mix between the water particles is a simple explanation.

Experiments with particles

Introduction

When materials are added together, they may acquire new properties. When a solid and a liquid are mixed, the solid may or may not dissolve. When two liquids are mixed they may become one liquid or stay separate. These experiments provide an opportunity to predict and then observe what happens.

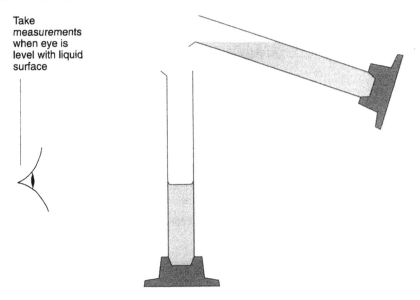

Take measurements when eye is level with liquid surface

What to record

Activity 1

Volume of peas/cm³	Volume of sand/cm³	Combined volume/cm³

Activity 2

Volume of alcohol/cm³	Volume of water/cm³	Combined volume /cm³

Activity 3

Initial volume of water/cm³	Final volume of salt solution/cm³

RS•C

What to do

Activity 1
1. Add approximately 25 cm^3 of dried peas and 25 cm^3 of sand to separate measuring cylinders. Accurately measure and record the volumes.
2. The contents of one cylinder is added to the other and shaken until the two substances are mixed together.
3. Place the measuring cylinder on the bench and gently shake from side to side to allow the mixture to settle.
4. Read the combined volume.

Activity 2
1. Add approximately 25 cm^3 of ethanol (**Highly flammable**) and 25 cm^3 of water to separate measuring cylinders. Accurately measure and record the volumes.
2. The contents of one cylinder is added to the other and shaken from side to side for 15–30 seconds until the two substances are mixed together and then left to stand for one minute.
3. Read the combined volume.

Activity 3
1. To the measuring cylinder add approximately 75 cm^3 of water. Accurately measure and record the volume.
2. Spatulas of salt should then be added one at a time until the salt begins to be left at the bottom of the cylinder, despite continued stirring.
3. The volume reading on the side of the cylinder should again be recorded.

Safety

Wear eye protection.

Questions

1. What is the similarity between the first two activities?
2. What is an explanation for the result in the last activity?

37. Particles in motion?

Topic

Particulate nature of matter, gases.

Timing

30 min.

Description

Students produce carbon dioxide by reacting calcium carbonate with hydrochloric acid. They then check to see if diffusion occurs by both holding the test-tube of carbon dioxide over a test-tube of air and vice-versa.

Apparatus and equipment (per group)

- Three test-tubes
- Cork
- Delivery tube and bung.

Chemicals (per group)

- Limewater 0.02 mol dm^{-3}
- Calcium carbonate
- Hydrochloric acid 0.5 mol dm^{-3}.

Teaching tips

This experiment provides a good introduction, one suggestion is to show a demonstration of Brownian motion using a smoke cell after this experiment.

Background theory

Solids, liquids and gases consist of minute particles. If this were not the case, they would not mix so easily. This is not proof of a particulate theory, but the experiment does suggest that the particles in the gas must be in motion to spread through the air in the containers.

Safety

Wear eye protection.

Answers

1. All the test-tubes contained carbon dioxide; the gases always diffuse and mix.
2. Carbon dioxide is denser than air.
3. Yes; both tubes should give cloudy limewater suggesting the gases in the two tubes mixed. Some of the heavier carbon dioxide molecules moved upwards into the test-tube containing air.

Particles in motion?

Introduction

These two activities suggest that particles in a gas are in motion.

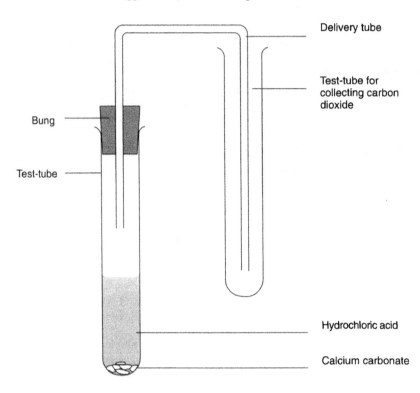

What to do

1. Set up the apparatus as shown in the diagram.
2. Put a spatula measure of calcium carbonate into the first test-tube.
3. Add 10 cm^3 of hydrochloric acid and quickly replace the bung and delivery tube. Ensure the delivery tube reaches almost to the bottom of the second test-tube.
4. Allow the gas to pass into the second test-tube for about one minute, then remove the delivery tube and cork the test-tube.
5. Hold the test-tube upside down over a similar test-tube containing air.
6. Remove the cork and place the tubes mouth to mouth.
7. After 5 min, cork both tubes and test the contents for carbon dioxide (swirl a little limewater round in the test-tube). Write down what happens in both tubes.
8. Repeat this experiment but this time at step 5 hold the test-tube of air upside down over a test-tube of carbon dioxide.

Safety

Wear eye protection.

Questions

1. Which of the four test-tubes contained carbon dioxide at the end of the experiment?
2. Is air or carbon dioxide more dense?
3. Does this experiment support the idea that the particles of a gas are in motion? Give your reasons.

38. Making a pH indicator

Topic

Acids and bases, indicators.

Timing

30 min.

Description

Students make a pH indicator from red cabbage.

Apparatus and equipment (per group)

- 250 cm^3 Beaker
- Tripod
- Gauze
- Bunsen burner
- Test-tube rack
- Three test-tubes.

Chemicals (per group)

- Red cabbage (3–4 small pieces)
- Dilute hydrochloric acid   0.5 mol dm^{-3}
- Calcium hydroxide solution   0.4 mol dm^{-3}.

Teaching tips

First, supervise carefully as students boil their cabbage for 10 min. Bunsen burners can then be switched off. Beakers can then be lifted carefully off the tripods.

Background theory

Students should understand the concept of an indicator.

Safety

Wear eye protection.

Answers

1. Green/yellow.
2. Purple/blue.
3. Red.

Making a pH indicator

Introduction

A pH indicator is a substance that has a different colour when added to acid or alkali. In this experiment a pH indicator is made from red cabbage.

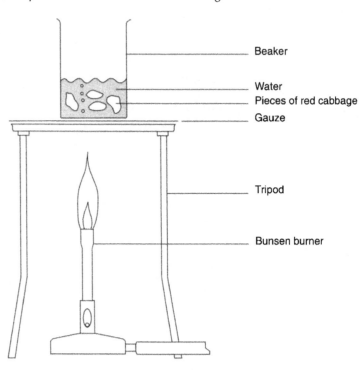

What to record

Record the colour of the cabbage indicator in the three solutions.

What to do

1. Boil about 100 cm^3 of tap water in a beaker.
2. Add three to four pieces of red cabbage to the boiling water.
3. Boil for about 5 min. The water should have turned blue or green.
4. Turn off the Bunsen burner and allow the beaker to cool for a few minutes.
5. Place three test-tubes in a rack. Half fill one with alkali, one with acid and one with deionised water.
6. Decant approximately 2–3 cm height of cabbage solution into each test-tube.

Safety

Wear eye protection.

Questions

1. What colour is the cabbage indicator when neutral?
2. What colour is the cabbage indicator when alkali?
3. What colour is the cabbage indicator when acid?

39. The reaction between a metal oxide and a dilute acid

Topic

Acids and bases.

Timing

30 min.

Description

Copper(II) oxide is dissolved in hot dilute sulfuric acid to give copper(II) sulfate.

Apparatus and equipment (per group)

- ▼ Tripod
- ▼ Gauze
- ▼ Filter paper
- ▼ Bunsen burner
- ▼ 100 cm^3 Conical flask
- ▼ 100 cm^3 Beaker
- ▼ Filter funnel
- ▼ Spatula.

Chemicals (per group)

- ▼ Sulfuric acid 0.4 mol dm^{-3}
- ▼ Three spatula measures copper(II) oxide (**Harmful**)

Teaching tips

Whatman qualitative No. 1 filter papers work well for this experiment. The teacher may need to remind students how to fold filter paper. A fluted filter paper allows faster filtration.

Background theory

Acid + base → salt + water.

Safety

Wear eye protection. Keep test-tubes facing away while heating.

Answers

1. Unreacted copper(II) oxide.
2. Copper(II) sulfate.
3. $CuO(s) + H_2SO_4(aq) \rightarrow CuSO_4(aq) + H_2O(l)$
4. The water can be evaporated, the solid copper(II) sulfate crystallises.

The reaction between a metal oxide and a dilute acid

Introduction

Many metal oxides react with dilute acid. In this experiment copper(II) oxide is reacted with dilute sulfuric acid.

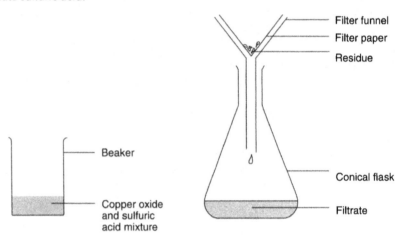

What to do

1. Pour about 20 cm^3 of dilute sulfuric acid into a beaker.
2. Heat on a tripod and gauze using a Bunsen burner until the acid just boils.
3. Add copper(II) oxide to the hot acid, a spatula measure at a time and stir after each addition, continue until no more dissolves.
4. Filter the mixture, while still hot, into a conical flask.

Safety

Wear eye protection.

Questions

1. What is the residue on the filter paper?
2. What does the filtrate contain?
3. Write an equation for the reaction between solid copper(II) oxide and dilute sulfuric acid.
4. How could the solid product be isolated from the filtrate?

40. Disappearing ink

Topic

Acids and bases.

Timing

30 min.

Description

Students produce a solution in which the colour disappears due to an acid/base reaction.

Apparatus and equipment (per group)

- ▼ 100 cm^3 Beaker
- ▼ 10 cm^3 Measuring cylinder
- ▼ Small paint brush to test the ink.

Chemicals (per group)

- ▼ Ethanol (**Highly flammable**)
- ▼ Sodium hydroxide 0.4 mol dm^{-3} (**Irritant**)
- ▼ Thymolphthalein solution (50 per cent water, 50 per cent ethanol) (**Highly flammable**)

Teaching tips

This ink is the same as those sold in trick and joke shops. The amount of indicator can be adjusted to give a deep blue colour. The compound produced, Na_2CO_3, is commonly called washing soda.

Background theory

Sodium hydroxide reacts with carbon dioxide in the air to form sodium carbonate.

$2NaOH(aq) + CO_2(g) \rightarrow Na_2CO_3(aq) + H_2O(l)$

Sodium carbonate is less basic than sodium hydroxide and causes the indicator to change from blue to colourless. The colourless range for thymolphthalein is below pH 9.3. The blue range is above pH 10.5 and the colour change takes place between these two. The alcohol evaporates and leaves a clear and colourless residue.

Safety

Wear eye protection.

Answers

1. Carbon dioxide.
2. Sodium hydroxide + carbon dioxide → sodium carbonate + water
3. $2NaOH + CO_2 \rightarrow Na_2CO_3 + H_2O$

Disappearing ink

Introduction

A blue liquid is made. This liquid is tested on a white page, it leaves a blue ink spot. In a few seconds, the blue spot disappears.

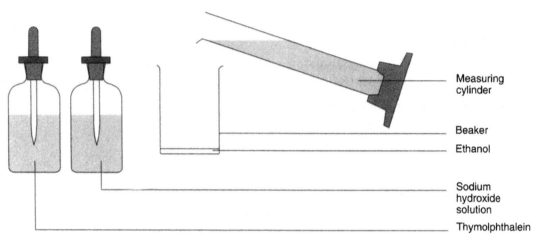

What to do

1. Place 10 cm^3 of ethanol (**Highly flammable**) in a small beaker.
2. Add a few drops of thymolphthalein (**Highly flammable**) indicator solution.
3. Add just enough NaOH solution (**Irritant**), dropwise, to produce a deep blue colour in the solution.
4. Using a small paint brush test the 'disappearing ink' on a white page.

Safety

Wear eye protection. Care with teat pipettes that contain sodium hydroxide.

Questions

The colour change occurs because sodium hydroxide reacts with a gas in the air.

1. Which gas in the air causes this colour change?
2. Write a word equation for the reaction.
3. Write a formula equation for the reaction.

41. Testing for enzymes

Topic

Catalysis, rates of reaction, enzymes.

Timing

30 mins.

Description

Students use the enzymes in liver, potato and celery to catalyse the production of water and oxygen from hydrogen peroxide.

Apparatus and equipment (per group)

- ▼ 100 cm^3 Conical flask
- ▼ 25 cm^3 Measuring cylinder.

Chemicals (per group)

- ▼ Hydrogen peroxide solution (5 volume)
- ▼ Liver (small piece)
- ▼ Potato (small piece)
- ▼ Celery (small piece).

Teaching tips

It may be useful to discuss how to decide which reaction is fastest. Teachers should be sensitive to the needs of vegetarian students. Manganese dioxide, copper oxide and calcium carbonate can also be tested as catalysts to illustrate biological and chemical catalysts. Cooked liver (well done) can also be used. This will not work as well, illustrating the fact that enzymes can be denatured.

A temperature sensor or a pressure sensor attached to a computer can be used to monitor the progress of this reaction. The technique can be used to provide a more graphic demonstration or indeed as the basis for an investigation into rates of reactions.

Background theory

Students should understand the concept of an enzyme as a biological catalyst.

Safety

Wear eye protection. Provide a bucket to collect discarded liver, do not allow it to collect in sinks.

Answers

1. Oxygen is produced.
2. It can relight a glowing splint.
3. The liver.
4. Hydrogen peroxide → oxygen + water
5. Rate of mass loss. Use a gas syringe, amount of fizz, or any other suitable method.

Testing for enzymes

Introduction

Enzymes are biological catalysts, they increase the speed of a chemical reaction. They are large protein molecules and these enzymes are very specific to certain reactions. Hydrogen peroxide decomposes slowly in light to produce oxygen and water. There is an enzyme called catalase that can speed up (catalyse) this reaction.

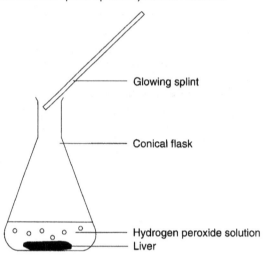

What to record

What do you see? What gas is produced, and which enzyme source makes the most effective catalyst?

What to do

1. Using a measuring cylinder, put 25 cm^3 of hydrogen peroxide solution into a conical flask.
2. Add a small piece of liver.
3. Test the gas given off with a glowing splint.
4. Dispose of this mixture, including the liver, into a bucket, and put another 25 cm^3 of hydrogen peroxide solution in the flask.
5. Add a small piece of potato.
6. Test the gas given off with a glowing splint.
7. Repeat this experiment with a piece of celery instead of potato.

Safety

Wear eye protection.

Questions

1. Which gas is produced in this reaction?
2. What is the test for this gas?
3. Which enzyme source produces the fastest reaction (liver, potato or celery)?
4. Write a word equation for this reaction.
5. How could the rate of gas production be measured?

42. Testing water hardness

Topic

Water hardness.

Timing

60 min.

Description

Students test the hardness of various types of water by finding the volume of soap required to form a permanent lather.

Apparatus and equipment (per group)

- 100 cm^3 Conical flask
- Bung to fit the conical flask
- Burette and burette stand
- 10 cm^3 Measuring cylinder.

Chemicals (per group)

Access to:

- Soap solution (**Highly flammable**)
- Deionised water (labelled Rainwater)
- Permanent hard water (labelled Seawater boiled and cooled)
- Temporary hard water (labelled Temporary hard water)
- Deionised water (labelled Temporary hard water boiled and cooled)
- A mixture of temporary and permanent hard water labelled (Seawater).

For soap solution:
Add 10 g of soft soap (Fisons ('soft soap')/Griffin & George (Soap soft green BP) to 100 cm^3 water. Warm gently. Make up to 1 dm^3 with a 50/50 solution of ethanol (**Highly flammable**) and water. The soap does precipitate out after a while. Alternatively purchase Wanklyn's standard soap solution.

For temporary hard water:
Bubble CO_2 through a mixture of 250 cm^3 limewater/ 750 cm^3 deionised water for about 20 min or until the cloudy precipitate disappears completely. Filter.

For permanent hard water:
Add hydrated calcium sulfate to deionised water. Allow to stand, then decant off the liquid.

For temporary and permanent hard water:
Use a 50 per cent mixture of hard and soft water.

Teaching tips

If burettes are used with younger students it is advisable to provide them ready clamped and filled to the zero mark.

RS•C

Background theory

Theory behind the different types of hardness, and how hardness is removed.

Safety

Wear eye protection.

Answers

1. Soft.
2. Hard.
3. Both.

Testing water hardness

Introduction

Tap water in some parts of the country is very pure and is said to be 'soft'. It easily makes a lather with soap. Water from other parts may contain various dissolved impurities and is described as 'hard' water. Temporary hardness may be removed by boiling, but permanent hardness survives the boiling process.

In this practical activity, water hardness can be measured by finding out the volume of soap solution required to form a permanent lather with a known volume of water.

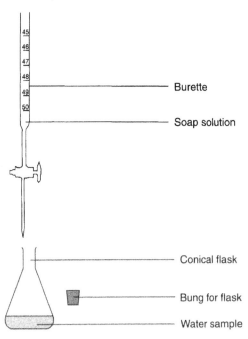

What to record

Record the volume of soap needed to produce a permanent lather with each type of water. Note any difference between the appearance of the samples after the addition of soap solution.

Water type	Volume of soap required a to produce permanent lather /cm^3
Rainwater	
Seawater	
Temporary hard water	
Seawater, boiled then cooled (permanent hard water)	
Temporary hard water, boiled then cooled	

RS•C

What to do

1. Collect a conical flask and bung. Check the bung is a good fit.
2. Measure 10 cm^3 of water sample into a conical flask using a measuring cylinder.
3. Using the burette add 1 cm^3 of soap solution to the water. Stopper the flask and shake vigorously. If no lather is produced, add another 1 cm^3 of soap solution. Continue in this way until a permanent lather (one that lasts for 30 seconds) is obtained. Record the volume of soap solution needed to produce a permanent lather.
4. Repeat this procedure for the other water samples.

Safety

Wear eye protection.

Questions

1. Is the rainwater hard or soft?
2. Is seawater hard or soft?
3. Does seawater contain temporary hardness, permanent hardness or both?

43. A chemical test for water

Topic

Indicators, qualitative analysis.

Timing

30–45 min.

Description

A cobalt(II) chloride solution is mixed with an equal amount of sodium chloride solution. Filter paper is dipped into the mixture and dried to make cobalt(II) chloride paper.

Apparatus and equipment (per group)

- One small beaker
- Filter paper
- Bunsen burner
- Gauze
- 250 cm^3 Beaker.

Chemicals (per group)

- Cobalt(II) chloride solution
- Sodium chloride solution.

(Cobalt(II) chloride solution is made by adding deionised water to 12 g of $CoCl_2.H_2O$ to make 100 cm^3 of solution. Sodium chloride solution is made by adding deionised water to 6 g of NaCl to make 100 cm^3 of solution.)

Teaching tips

1. A small amount of solution goes a long way. Students can share the same beaker.
2. The concentrations of the solutions are not too important, as long as the pink colour is noticeable on the paper.
3. A hair dryer can be used to dry the papers (these should not be bought in from home). They can also be carefully dried over a Bunsen burner.
4. Students could place the blue paper in several places around the school and monitor humidities.

Background theory

Cobalt atoms in salts are positive ions with a 2+ charge. They attract negative particles such as Cl^- and the lone pair on the oxygen in water molecules. When most of the negative species around the cobalt ion are water molecules, the ion absorbs light so that it appears pink. When the paper is dry, the water molecules have been driven off. The negative chloride ion sticks to the positive cobalt ions, and the cobalt appears blue. The cobalt complexes with the water and causes the electrons in the cobalt to absorb at different energies. These different energies result in different colours being absorbed by the cobalt.

$$Co(H_2O)_6^{2+}(aq) + 4Cl^-(aq) \rightarrow Co(Cl_4)^{2-}(aq) + 6H_2O(l)$$
$$\text{pink} \qquad\qquad\qquad\qquad \text{blue}$$

Safety

Wear eye protection and protective gloves when preparing solutions, as $CoCl_2$ solid is a possible sensitiser.

Answers

1. Sodium chloride picks up moisture from the air.
2. Indefinitely, as long as the water does not wash the salts off the paper.
3. It can be used to monitor humidity or identify the presence of water anywhere. It can also be used to detect water in petrol samples and in certain chemicals.

A chemical test for water

Introduction

Some chemicals change colour when water is added to them. Some coloured chemicals owe their colour to the water molecules that are associated with them. Cobalt(II) chloride is one colour when dry and another colour when damp. In this experiment these colours are identified.

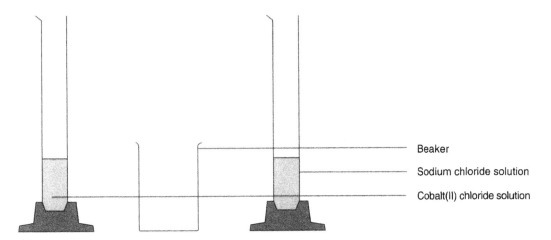

Beaker
Sodium chloride solution
Cobalt(II) chloride solution

What to do

1. Add 4 cm^3 of cobalt(II) chloride solution to a small beaker.
2. Add 4 cm^3 of salt solution.
3. Dip half of the filter paper into the solution, using tongs.
4. Boil a beaker of water and carefully lift the beaker off the tripod onto the bench.
5. Allow the paper to dry. Wrap it around the beaker of hot water to speed up drying.
6. Observe differences in colour between the wet and the dry cobalt(II) chloride paper.
7. Place the dry cobalt(II) chloride paper near an open window on a humid day and observe what happens.

Safety

Wear eye protection.

Avoid contact with the cobalt(II) chloride paper and the cobalt(II) chloride solution. Cobalt(II) chloride is toxic.

Questions

1. For what other purposes might the sodium chloride be on the paper, in addition to supplying more chloride ions?
2. How many times can the cobalt(II) chloride paper cycle between colours?
3. Suggest a practical application for the cobalt(II) chloride paper.

RS•C

44. Forming glass

Topic

Making glass, ceramics.

Timing

60 min.

Description

Students make glass and then colour some glass using transition metal oxides.

Apparatus and equipment (per group)

- ▼ Access to a balance (±0.1 g)
- ▼ Paper clip
- ▼ Bunsen burner
- ▼ Tripod
- ▼ Pipe clay triangle
- ▼ Crucible
- ▼ Heat proof mat
- ▼ Tongs
- ▼ Boiling tube and bung.

Chemicals (per group)

Access to:

- ▼ Copper(II) oxide (**Harmful**)
- ▼ Cobalt(II) oxide (**May cause sensitisation, harmful**)
- ▼ Manganese(II) oxide
- ▼ Chromium(III) oxide (Not chromium(VI) oxide)
- ▼ Lead(II) oxide (**Toxic, may cause harm to unborn children and there is a danger of cumulative effects. Harmful if swallowed.**)
- ▼ Boric acid
- ▼ Zinc oxide.

Teaching tips

The solids need to be thoroughly mixed. Students can put all three solids in a boiling tube, use the bung and shake the contents to ensure thorough mixing. Stir the mixture with an unfolded paper clip. The crucibles need to be dedicated to this experiment. This practical is suitable for teaching how to use balances and how to handle hot apparatus. The glass produced is very brittle and difficult to keep. It is better to have a number of balances available.

Background theory

Very little required. Ceramics and their properties. Practical skills are more important.

Safety

Wear eye protection. Avoid raising lead(II) oxide dust.

Forming glass

Introduction

The aim of this experiment is first to make some glass and secondly to make some coloured glass by adding other compounds to the molten glass mixture.

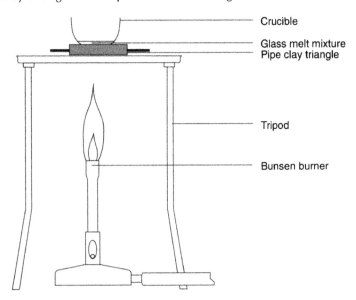

What to record

Record the colour of the glass produced when a speck of a particular oxide is added.

What to do

1. Weigh 6.5 g of lead(II) oxide, 3.5 g of boric acid and 0.5 g of zinc oxide. Mix thoroughly in the boiling tube.
2. Pour the mixture into a crucible and place it on a pipe clay triangle.
3. Heat strongly until it becomes molten and runny. Using tongs pour one or two drops of the molten glass onto your heatproof mat.
4. Allow the beads to cool for 5 min and then examine them.
5. Using an unfolded paper clip pick up a tiny amount of transition metal oxide and drop it into the remaining molten mixture. Stir in the powder using the paper clip.

Do not add too much powder or you will produce a very dark piece of glass.

Safety

Wear eye protection. Care must be taken with the lead(II) oxide, as it is toxic. Avoid raising the dust. Some of the other chemicals are harmful.

45. Thermometric titration

Topic

Titration, neutralisation, energy changes and salt formation.

Timing

60 min.

Description

Students titrate sodium hydroxide with hydrochloric acid. The temperature is measured each time a portion of acid is added. The highest temperature indicates the endpoint of the reaction and this is used to estimate the molarity of the hydrochloric acid.

Apparatus and equipment (per group)

- ▼ Thermometer (0–100 °C) supported at its upper end (in a cork) in a clampstand
- ▼ Polystyrene cup
- ▼ Burette and burette stand
- ▼ Measuring cylinder (>15 cm^3).

Chemicals (per group)

- ▼ Hydrochloric acid   1.5 mol dm^{-3} (concentration not indicated on bottle)
- ▼ Sodium hydroxide   2 mol dm^{-3} (**Corrosive**) (concentration indicated on bottle).

Teaching tips

The main concern in this experiment is the heat loss. With more able or older students, it is possible to discuss the extrapolation of the cooling curve. To reinforce the point an indicator could also be used to show that the end point really was at the highest temperature. If possible a lid should be used.

Teachers may wish to develop the theory of titration further using this experiment.

A temperature sensor attached to a computer can be used here in place of the thermometer. Data logging software will show the temperature change as acid is added to alkali. Adding a pH sensor to the experiment helps to show that after the equivalence point is reached, the temperature stops rising.

Background theory

- ▼ Neutralisation reactions
- ▼ Energy changes in chemical reactions
- ▼ Moles and solutions
- ▼ Graph plotting skills.

Safety

Wear eye protection.

Thermometric titration

Introduction

The aim of this experiment is to measure the maximum temperature reached during the reaction between hydrochloric acid and sodium hydroxide solution. The solutions of acid and alkali do not have the same concentration. The volumes that have reacted at the highest temperature reached, represent the 'end point' of the titration.

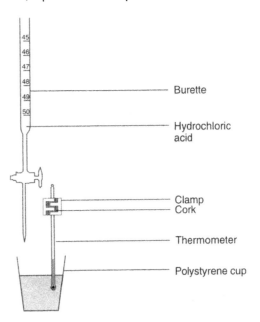

What to record

Record your results in a table.

Volume of acid added/cm^3	Temperature/ °C

What to do

1. Using a measuring cylinder, place 15 cm^3 of sodium hydroxide (**Corrosive**) into the polystyrene cup and measure the temperature.
2. Using the burette add a small portion (3–5 cm^3) of dilute hydrochloric acid to the solution in the polystyrene cup. Swirl the solution and measure the highest temperature reached.
3. Immediately add a second small portion of the dilute hydrochloric acid, swirl and again measure the highest temperature.
4. Continue in this way until there are enough readings to decide the highest temperature for the experiment.

Safety

Wear eye protection.

Questions

1. What is the highest temperature reached in this reaction?
2. Draw a graph of your results.

46. The formation of metal crystals

Topic

Electrolysis, growing metal crystals.

Timing

20 min.

Description

Students carry out the electrolysis of tin(II) chloride solution. Spectacular crystals of tin are observed.

Apparatus and equipment (per group)

- ▼ Power pack (2–4 V)
- ▼ Two leads
- ▼ Two crocodile clips
- ▼ Steel nail (>2 cm)
- ▼ Carbon electrode
- ▼ 100 cm^3 Beaker
- ▼ Plastic electrode holder.

Chemicals (per group)

Tin(II) chloride solution (40 cm^3) approximately 0.5 mol dm^{-3} (**Irritant**)
113 g of tin(II) chloride dihydrate (**Irritant**) (formula mass 225.6 g) is dissolved in 200 cm^3 of concentrated hydrochloric acid (**Corrosive**), a few pieces of metallic tin are added, and then the solution is made up to 1 dm^3 with deionised water. The solution is irritant.

Teaching tips

The best voltage is between 2–4 V. Ensure the nail is the cathode.

Background theory

Students should understand what ions are, and the concept of electrolysis.

Safety

Wear eye protection. Ensure the room is well ventilated or carry out the experiment in a fume cupboard. Do not run the current for too long as toxic chlorine gas is produced.

Answers

1. Chlorine
2. Tin cans
3. $Sn^{2+} + 2e^- \rightarrow Sn$

The formation of metal crystals

Introduction

Metal crystals can be grown using several methods – eg displacing one metal by another from a salt solution, by cooling a liquefied metal and by electrolysing a salt solution. This experiment illustrates the electrolysis of tin(II) chloride solution.

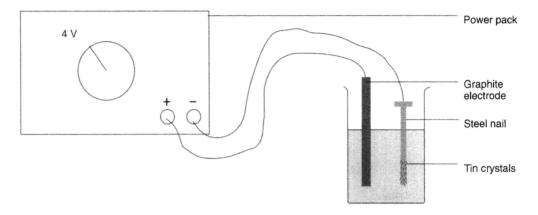

What to record

What was done and what was observed.

Note your observations at each electrode.

What to do

1. Set up an electrolysis experiment as shown in the diagram.
2. Ensure the nail is connected as the cathode (-ve terminal) and the carbon rod as the anode (+ve terminal).
3. Set the voltage to 4 V
4. Observe the formation of tin crystals on the nail cathode.

Safety

Wear eye protection. Do not inhale the fumes produced. Do not electrolyse for too long. The tin(II) chloride is dissolved in strong acid (**Irritant solution**).

Questions

1. What is the product formed at the carbon anode?
2. Give one common household use of steel electroplated with tin.
3. Write an ionic equation for the reaction at the cathode.

47. Formation of a salt which is insoluble in water

Topic

Insoluble salt formation, precipitation.

Timing

20 min.

Description

Students react magnesium sulfate and sodium carbonate to form magnesium carbonate, which is insoluble in water.

Apparatus and equipment (per group)

- Two 250 cm^3 conical flasks
- Filter funnel
- Filter paper
- A 25 cm^3 or larger measuring cylinder.

Chemicals (per group)

Magnesium sulfate solution	1 mol dm^{-3}
Sodium carbonate solution	1 mol dm^{-3}.

Teaching tips

Filtration can take a while, so students can begin to clear up.

Background theory

What is a salt? Formation of salts.

Safety

Wear eye protection.

Answers

1. Magnesium sulfate + sodium carbonate → magnesium carbonate + sodium sulfate
2. $MgSO_4(aq) + Na_2CO_3(aq) \rightarrow MgCO_3(s) + Na_2SO_4(aq)$
3. $Mg^{2+}(aq) + CO_3^{2-}(aq) \rightarrow MgCO_3(s)$
4. The spectator ions are the ions left in solution. These are called spectator ions because they do not play a part in the reaction.

Formation of a salt which is insoluble in water

Introduction

When solutions of two soluble salts are mixed, a solid may form. The solid is called a precipitate, and the reaction is called a precipitation reaction. Precipitation reactions are used to make insoluble salts.

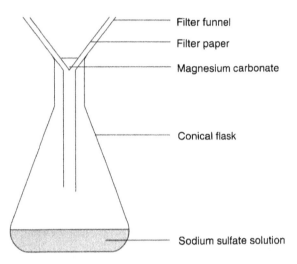

What to record

What was done and what was observed.

What to do

1. Mix 25 cm^3 of magnesium sulfate solution and 25 cm^3 of sodium carbonate solution in a conical flask.
2. Filter the mixture to remove the spectator ions.
3. Remove the filter paper with the magnesium carbonate and leave to dry.

Safety

Wear eye protection.

Questions

1. Write a word equation for this reaction.
2. Write a formula equation for this reaction.
3. Write an ionic equation for this reaction.
4. What is meant by spectator ions?

RS•C

48. Titration of sodium hydroxide with hydrochloric acid

Topic

Acids and alkalis, neutralisation, formation of a soluble salt.

Timing

45 min.

Description

Students find the amount of acid required to neutralise 25 cm^3 of sodium hydroxide. These volumes are reacted and sodium chloride is then crystallised.

Apparatus and equipment (per group)

- ▼ 250 cm^3 Conical flask
- ▼ 50 cm^3 or 100 cm^3 Burette
- ▼ Burette stand
- ▼ One 25 cm^3 or larger measuring cylinder.

Chemicals (per group)

- ▼ Hydrochloric acid 1 mol dm^{-3}
- ▼ Sodium hydroxide 1 mol dm^{-3} (**Corrosive**)
- ▼ Methyl orange.

Teaching tips

Demonstrate how to fill a burette and run out to the zero mark.

With older groups, a pH sensor in the flask can monitor the pH change as the alkali is neutralised. Get the computer recording the pH sensor and allow the acid to drip at a steady rate as you swirl the flask. Data logging software will show the readings as a 'classic' strong acid-strong base titration curve where pH is on the y-axis and time (as a rough measure of the volume) is on the x-axis.

Background theory

Acids and alkali, salts.

Safety

Wear eye protection.

Answers

1. Salt.
2. Hydrochloric acid + sodium hydroxide → sodium chloride + water
3. HCl + NaOH → NaCl + H$_2$O
4. The original solution contained methyl orange.

Titration of sodium hydroxide with hydrochloric acid

Introduction

In this experiment sodium hydroxide is neutralised with hydrochloric acid to produce the soluble salt sodium chloride. This is then concentrated and crystallised in a crystallising dish.

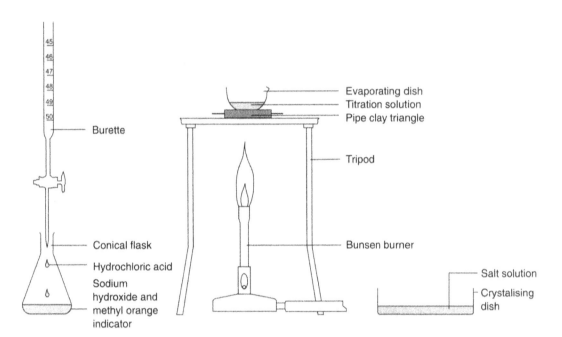

What to record

What was done and what was observed.

What to do

1. Add 25 cm^3 of sodium hydroxide solution (**Corrosive**) to a conical flask using a measuring cylinder and add a couple of drops of methyl orange indicator.
2. Fill the burette with hydrochloric acid and run through to the zero mark (use a funnel to fill the burette and a beaker to collect the excess acid).
3. Add the hydrochloric acid to the sodium hydroxide solution in small volumes swirling after each addition. Continue until the solution turns red and record this reading on the burette.
4. Carefully add this volume of fresh hydrochloric acid to another 25 cm^3 of sodium hydroxide solution to produce a neutral solution.
5. Reduce to about half the volume using an evaporating dish on a gauze over a Bunsen burner flame.
6. Leave to evaporate in a crystallising dish to produce a white crystalline solid.

RS•C

Safety

Wear eye protection.

Questions

1. What is the everyday name for sodium chloride?
2. This reaction is a specific example of the general reaction:
   Acid + alkali → salt + water.
   Write a word equation for this specific reaction.
3. Write a formula equation for this reaction.
4. Why must you use another 25 cm^3 of sodium hydroxide solution to make *pure* sodium chloride?

49. The properties of ammonia

Topic

Acids and alkalis, solubility, ammonia.

Timing

45 min.

Description

Students produce ammonia and carry out the fountain experiment themselves.

Apparatus and equipment (per group)

- ▼ Two boiling tubes
- ▼ Delivery tube
- ▼ Bung with jet, rubber tube and clip
- ▼ 250 cm^3 Beaker
- ▼ Retort stand
- ▼ Boss
- ▼ Clamp
- ▼ Bunsen burner.

Chemicals (per group)

- ▼ Universal Indicator solution
- ▼ Red litmus paper
- ▼ Universal Indicator paper
- ▼ Calcium hydroxide
- ▼ Ammonium chloride (**Harmful**).

Teaching tips

Slope the reaction tube as shown in the diagram. This ensures water drains away from the reaction. A selection of indicators could be provided for the fountain experiment.

Background theory

Acids and alkalis.

Safety

Wear eye protection. Carry out the experiment in a well ventilated room because ammonia gas is toxic.

Answers

1. Ammonia is a colourless gas with a strong, choking smell. It does not burn and puts out a lighted splint. It is less dense than air. It is alkali to litmus and Universal Indicator. It is very soluble in water, as shown by the fountain experiment. It reacts with hydrochloric acid to form ammonium chloride.

2. Ammonium chloride + calcium hydroxide → calcium chloride + water + ammonia

The properties of ammonia

Introduction

In this experiment, ammonia is produced and collected. The gas is tested and the solubility in water is illustrated by the use of a fountain experiment.

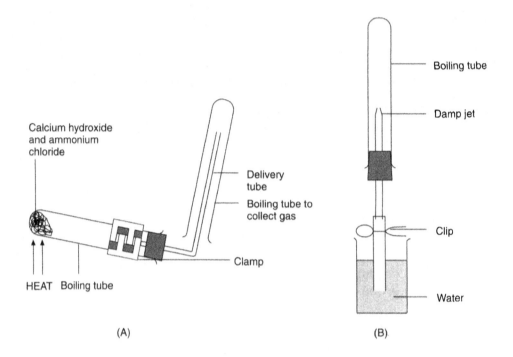

What to record

Observe what happens at each stage. Record the results in the table.

Test	Observations
Heating the mixture	
Lighted splint test	
Damp red litmus paper	
Damp Universal Indicator paper	
Hydrochloric acid bottle stopper	
Opening the clip	

What to do

1. Add two spatulas of calcium hydroxide and two spatulas of ammonium chloride to a boiling tube and mix them.
2. Set up the apparatus as shown in the diagram (A). Warm gently.

3. Test the gas produced with a lighted splint
4. Test the gas with damp red litmus paper
5. Test the gas with damp Universal Indicator paper
6. Test the gas with a stopper from a hydrochloric acid bottle.
7. Fill a dry boiling tube with the gas by heating for several minutes. (The tube *must* be dry.)
8. Fit the tube quickly with a bung carrying a damp glass jet.
9. Set up the apparatus as shown in diagram (B).
10. Open the clip.

Safety

Wear eye protection. Take care not to inhale the ammonia produced. Work in a well ventilated area.

Questions

1. Fill in the missing gaps: Ammonia is a _____ gas with a strong, choking _____. It does not _____ and puts out a lighted _____. It is ____ _____ than air. It is _____ to litmus and Universal Indicator. It is very _____ in water, as shown by the _____ experiment. It reacts with hydrochloric acid to form _____ chloride.

2. Complete the following word equations:
Ammonium chloride + calcium hydroxide → _____ + water + _____

RS•C

50. Causes of rusting

Topic

Oxidation, rusting.

Timing

2 lessons (minimum).

Description

Students set up an experiment to find the conditions required for rusting to occur. Conditions tested are just air (oxygen), or just water, or both air and water.

Apparatus and equipment (per group)

- ▼ Three test-tubes
- ▼ Cotton wool
- ▼ Three steel nails
- ▼ Test-tube rack
- ▼ Two rubber bungs.

Chemicals (per group)

- ▼ Calcium chloride (anhydrous granules) (**Irritant**)
- ▼ Cooking oil
- ▼ Deionised water
- ▼ Boiled deionised water (15 min boil).

Teaching tips

This is an effective experiment to introduce the methods used to prevent rusting.

Background theory

Students should understand that rusting is the oxidation of iron to form iron oxide.

Safety

Wear eye protection.

Answers

1. Water is boiled to remove the dissolved oxygen. Oil is added to prevent the atmospheric oxygen dissolving.
2. The presence of air and water.
3. Any method that prevents the access of air and water to the iron.

Causes of rusting

Introduction

Rusting of iron and steel is a commonly occurring process with which we are all familiar. This experiment investigates the conditions needed for rusting to occur.

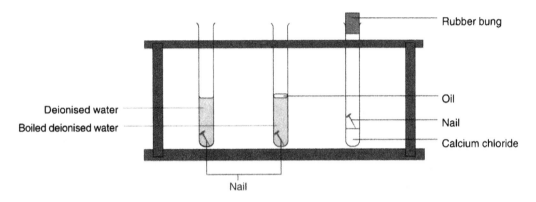

What to record

Observe what happens at each stage. Complete the table.

Tube number	Conditions	Observations when re-examined
1		
2		
3		

What to do

1. Place a clean nail into a test-tube that contains a little deionised water.
2. Place a clean nail into a test-tube that contains a little boiled deionised water. Pour about 1 cm depth of oil onto the surface.
3. Place about 2 cm depth of anhydrous calcium chloride granules into a test-tube, add a nail on top and place a bung on the tube.
4. Leave the tubes for more than three days and then re-examine the nails.

Safety

Wear eye protection.

Questions

1. Explain why the water is boiled, and oil added in tube 2.
2. What conditions are required for rust to form?
3. Suggest another method to prevent rust formation.

51. Reactions of calcium carbonate

Topic

Common reactions of chalk and limestone.

Timing

40 min.

Description

Students heat calcium carbonate to form calcium oxide. This is dissolved in water to form calcium hydroxide (limewater). Students then blow bubbles through this to form a cloudy suspension of calcium carbonate.

Apparatus and equipment (per group)

- Tripod
- Gauze
- Bunsen burner
- Tongs
- Two boiling tubes
- Drinking straw
- Universal Indicator solution
- Dropping pipette. Use the type of teat pipette (usually fitted to Universal Indicator bottles) that does not allow squirting. – eg Griffin.
- Filter paper
- Filter funnel.

Chemicals (per group)

- Calcium carbonate (small sample of chalk).

Teaching tips

A sample of soft chalk (calcium carbonate) reacts better than a marble chip. Blackboard chalk is not always calcium carbonate.

Background theory

A reminder about limewater and its reaction with carbon dioxide is advisable.

Safety

Wear eye protection. Remind students not to suck the calcium hydroxide into their mouths.

Answers

1. Calcium carbonate → calcium oxide + carbon dioxide
2. Calcium oxide + water → calcium hydroxide
3. Calcium hydroxide + carbon dioxide → calcium carbonate + water
4. $CaCO_3 \rightarrow CaO + CO_2$
5. $CaO + H_2O \rightarrow Ca(OH)_2$
6. $Ca(OH)_2 + CO_2 \rightarrow CaCO_3 + H_2O$

Reactions with calcium carbonate

Introduction

Limestone and chalk are mainly calcium carbonate. In this experiment, calcium carbonate is heated to form calcium oxide. This is reacted with a few drops of water, and the resulting calcium hydroxide is dissolved in water. Carbon dioxide is bubbled through the water and the milky suspension of calcium carbonate characteristic of limewater and carbon dioxide is observed.

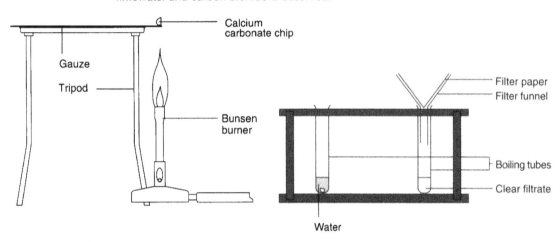

What to record

Observe what happens at each stage. Complete the table.

Method	Observation
Heat for 10 mins	
Add 2–3 drops of water	
Blow bubbles through solution	
Add Universal Indicator	

What to do

1. Set the chip of calcium carbonate, $CaCO_3$, on a gauze. If your gauze has a coated circle use the edge where there is no coating. Heat strongly for 10 minutes.
2. Let the chip cool and use tongs to move to a boiling tube. Add 2–3 drops of water with a dropper.
3. Add about 10 cm^3 more water to the solid. Then filter half the mixture into the other boiling tube.
4. Gently blow a few bubbles through the filtrate.
5. Test the remaining half with Universal Indicator solution.

Safety

Wear eye protection. Take care not to suck the limewater into your mouth.

Questions

Write word equations for the reactions that occur at the following stages.

1. Calcium carbonate is heated.
2. Water is added to the product.
3. Carbon dioxide is bubbled through limewater.

Write formula equations for the reactions that occur at these stages.

4. Calcium carbonate is heated.
5. Water is added to the product.
6. Carbon dioxide is bubbled through limewater.

52. To find the formula of hydrated copper(II) sulfate

Topic

Reversible reactions, formula mass, finding chemical formulae.

Timing

30 min.

Description

Students remove the water of crystallisation from hydrated copper(II) sulfate by heating. The change in mass is recorded, and hence the formula of the hydrated copper(II) sulfate is found.

Apparatus and equipment (per group)

- Tripod
- Pipe clay triangle
- Crucible
- Access to a balance (±0.01 g)
- Bunsen burner
- Tongs.

Chemicals (per group)

Hydrated copper(II) sulfate (**Harmful**).

Teaching tips

Remind students to zero the balance before use. Some students can do the relative molecular mass calculations while they are waiting for the balances to become free. A similar experiment can be done with $Na_2CO_3.xH_2O$.

A spreadsheet could be used to calculate the results for the class.

More able students could carry out a less structured calculation.

Background theory

Some reactions are easily reversible and others are not. An easily reversible reaction implies that strong bonds are not made. Concept of water of crystallisation. Mole calculations.

Safety

Wear eye protection.

Answers

1. $CuSO_4.5H_2O$

To find the formula of hydrated copper(II) sulfate

Introduction

In this experiment, the water of crystallisation is removed from hydrated copper(II) sulfate. The mass of water is found by weighing before and after heating. This information is used to find x in the formula: $CuSO_4.xH_2O$.

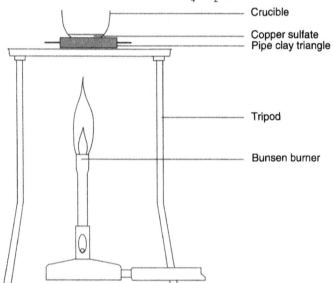

- Crucible
- Copper sulfate
- Pipe clay triangle
- Tripod
- Bunsen burner

What to record

Complete the table.

Relative atomic mass H=1, O=16, S=32, Cu=64

It is necessary to calculate the relative molecular mass of H_2O and $CuSO_4$.

What to do

1. Find the mass of your crucible.
2. Place 2–3 spatulas of blue copper(II) sulfate in the crucible and weigh.
3. Heat until the powder has gone completely white, but do not heat so strongly that it starts to blacken.
4. Allow to cool then reweigh.

(a)	Mass of crucible	–	g
(b)	Mass of crucible + blue copper(II) sulfate	–	g
(c)	Mass of crucible + white copper(II) sulfate	–	g
(d)	Mass of blue copper(II) sulfate	(b-a)	g
(e)	Mass of white copper(II) sulfate	(c-a)	g
(f)	Mass of water	(d-e)	
(g)	Moles of white copper(II) sulfate	e/RMM ($CuSO_4$)	
(h)	Moles of water	f/RMM(H_2O)	
(i)	Moles of water/Moles copper(II) sulfate	h/g	
(j)	Formula of hydrated copper(II) sulfate.	–	

Safety

Wear eye protection.

Question

1. What is the formula of hydrated copper(II) sulfate?

RS•C

53. Heating copper(II) sulfate

Topic

Reversible reactions.

Timing

30 mins.

Description

Students remove the water of crystallisation from hydrated copper(II) sulfate by heating. Condensing in a test-tube collects the water. The white anhydrous copper(II) sulfate can then be rehydrated, the blue colour returns.

Apparatus and equipment (per group)

- ▼ Two test-tubes
- ▼ Delivery tube (right angled)
- ▼ 250 cm^3 Beaker for cold water bath
- ▼ Bunsen burner
- ▼ Clamp and stand.

Chemicals (per group)

- ▼ Hydrated copper(II) sulfate (powdered)(**Harmful**).

Teaching tips

Ensure that the reaction test-tube is clamped at the end nearest the bung. Warn about and watch for 'suck back'. Demonstrate how to lift the entire clamp stand and apparatus.

Background theory

Some reactions are easily reversible and others are not. Concept of water of crystallisation.

Safety

Wear eye protection.

Answers

1. To cool and condense the steam.
2. (a) with water
   (b) without water
   (c) substance produced by a reaction
   (d) turned from vapour to liquid
   (e) a product is produced by the formation of new bonds.
3. The reaction can easily be reversed to produce the original reactants – *ie* hydrated copper(II) sulfate and heat.
4. The same amount of heat is required to form anhydrous copper(II) sulfate from the hydrated form, as that produced by the rehydration.

Heating copper(II) sulfate

Introduction

In this experiment the water of crystallisation is removed from hydrated blue copper(II) sulfate. After cooling the anhydrous copper(II) sulfate formed is then rehydrated with the same water.

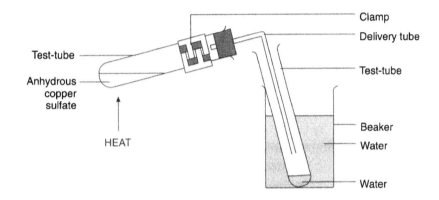

What to record

Record any observations when the water was poured back onto the white copper(II) sulfate.

What to do

1. Set up the apparatus as shown.
2. Heat the blue copper(II) sulfate until it has turned white.
3. Act quickly to prevent suck back. Lift the clamp stand so that the delivery tube does not reach into the water in the test-tube.
4. Allow the anhydrous copper(II) sulfate to cool.
5. Hold the tube containing anhydrous copper(II) sulfate in one hand and pour the condensed water onto the powder.

Safety

Wear eye protection.

Questions

1. Why is one test-tube placed in a beaker of cold water?
2. What do the following words mean (a) hydrated, (b) anhydrous, (c) product, (d) condensed and (e) reaction?
3. The reaction
   Hydrated copper(II) sulfate + heat $\rightleftharpoons$ anhydrous copper(II) sulfate + water
   is called a reversible reaction. Why?
4. Anhydrous copper(II) sulfate could be used as a fuel for heating ('just add water to get the heat'). Explain why it would not be a very economical fuel.

RS•C

54. The oxidation of hydrogen

Topic

Oxidation, combustion.

Timing

30 min.

Description

Students vary the proportions of hydrogen to air and decide which makes the loudest pop.

Apparatus and equipment (per group)

- 100 cm^3 Flask or boiling tube
- Delivery tube
- Water trough
- Marker pens (waterproof)
- Test-tubes with bungs (x4)
- Test-tube rack.

Chemicals (per group)

- Zinc
- Hydrochloric acid   2 mol dm^{-3} (**Irritant**)
- Copper(II) sulfate solution   0.8 mol dm^{-3}.

Teaching tips

If students are to make their own hydrogen then use 2 mol dm^{-3} hydrochloric acid (**Irritant**).

If the teacher generates the hydrogen: then use 4–5 mol dm^{-3} hydrochloric acid (**Irritant**), to increase the rate of production.

One suggestion is to have just one Bunsen burner on the teacher's bench, and students light their mixture from that.

A few drops of copper(II) sulfate solution catalyses the hydrogen production.

Background theory

This may be introduced early in a science course with the concept of a fire triangle and the fact that oxygen is needed for combustion. Later (14–16) this may be related to the proportion of reactants, lean burn in the internal combustion engine *etc*.

Safety

Wear eye protection. Collect all hydrogen generators before lighting any Bunsen burners.

Answers

1. The best pop is achieved with 20–40 per cent hydrogen.
2. Twice as many hydrogen as oxygen atoms (Teachers may decide not to discuss the formulae of hydrogen and oxygen molecules).
3. About 30 per cent hydrogen, 70 per cent air.

The oxidation of hydrogen

Introduction

In this experiment hydrogen is burnt with some air. The aim of the experiment is to find out how much air is needed to burn hydrogen most efficiently.

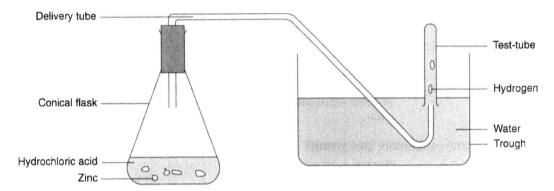

What to record

How loud/shrill is the pop for each mixture.

What to do

1. Mark one test-tube where it is a quarter full. Mark one at half full, another at three quarters full.
2. Set up the equipment as shown – a few drops of copper(II) sulfate speeds up the hydrogen production.
3. Using the mark as a guide, fill one test-tube a quarter, one a half, one three quarters and another completely full of water.
4. Invert the quarter filled tube in the trough to collect the hydrogen by displacing the water.
5. Put a bung in the test-tube.
6. Repeat this with the tubes half full, three quarters full and completely full of water.
7. Keep the 4 test tubes in the test-tube rack and clear away the hydrogen generator.
8. Using a lighted splint, carry out the 'pop' test on each of the tubes.

Safety

Wear eye protection.

Questions

1. Which mixture of hydrogen and air gives the most powerful explosion?
2. The formula of the product is H_2O, what is the ideal mixture of hydrogen and oxygen?
3. If the air is one-fifth oxygen calculate the ideal mixture of hydrogen to air?

RS•C

55. Investigating the reactivity of aluminium

Topic

Reactivity series.

Timing

30 min.

Description

Students add aluminium cooking foil to copper(II) sulfate solution and observe no reaction. Then sodium chloride is added and dissolved. A vigorous displacement reaction occurs and the solution gets very hot, aluminium dissolves and red copper is visible.

Apparatus and equipment (per group)

- ▼ 100 cm^3 Conical flask.

Chemicals (per group)

- ▼ Copper(II) sulfate solution   0.8 mol dm^{-3}
- ▼ Aluminium cooking foil (2 cm x 2 cm)
- ▼ Sodium chloride.

Teaching tips

Aluminium does not show its true reactivity until the oxide layer is disturbed. Sodium chloride disturbs this oxide layer. Scratches on the surface of the oxide layer allow chloride ions to react with aluminium, this effects the cohesiveness of the oxide layer. This allows reaction with the copper(II) sulfate. Remind students what copper looks like, so that they know what they are looking for.

Background theory

Reactive metals displace less reactive metals from solutions of their compounds.

Safety

Wear eye protection.

Answers

1. Aluminium appears less reactive than copper. The aluminium foil appears unable to displace copper from copper(II) sulfate solution.
2. Now aluminium is more reactive because it displaces copper.
   Aluminium + copper(II) sulfate → copper + aluminium sulfate
3. Scratches on the surface of the oxide layer allow chloride ions to react with aluminium, this effects the cohesiveness of the oxide layer. This allows a simple exchange reaction with the copper(II) sulfate. The protective oxide layer forms instantly the aluminium is exposed to the air.

Investigating the reactivity of aluminium

Introduction

This experiment illustrates the displacement of copper from copper(II) sulfate solution using aluminium foil.

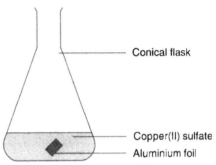

What to record

Write yes or no.

Observations	Before the sodium chloride is added	After the sodium chloride is added
Bubbles observed		
Colour changes		
Temperature change		
Copper observed		

What to do

1. Measure approximately 20 cm^3 of copper(II) sulfate solution into the conical flask.
2. Add a square of aluminium foil.
3. Look for signs of a reaction.
4. Add a spatula of sodium chloride and stir to dissolve.
5. Observe any changes. If nothing happens, add more sodium chloride. Has displacement of copper from copper(II) sulfate occurred?

Safety

Wear eye protection

Questions

1. Before the sodium chloride is added, does any reaction occur?
2. After adding sodium chloride, does the aluminium appear more or less reactive?
3. How does the salt affect this change?

RS•C

56. An oscillating reaction

Topic

Kinetics, redox reaction.

Timing

30 min.

Description

Students mix acidified sodium bromate solution with sodium bromide solution, then add propanedioic acid, ferroin and Photoflo (or some other surface active agent). The colour of the solution oscillates between red and blue.

Apparatus and equipment (per group)

- ▼ 100 cm^3 Beaker
- ▼ Petri dish.

Chemicals (per group)

- ▼ 6 cm^3 of solution A (Provide in a beaker with 10 cm^3 measuring cylinder) (**Irritant**)
- ▼ 0.5 cm^3 of solution B (Provide in a beaker with a 0–1 cm^3 syringe)
- ▼ 1.0 cm^3 of solution C (Provide in a beaker with a 0–1 cm^3 syringe)
- ▼ 1.0 cm^3 of ferroin (Provide in a beaker with a 0–1 cm^3 syringe)
- ▼ Photoflo (1 drop) (Provide in a beaker with a teat pipette)

Solution A: 5 g of sodium bromate and 2 cm^3 of concentrated sulfuric acid in 67 cm^3 of deionised water.

Solution B: 1 g of sodium bromide in 10 cm^3 of deionised water.

Solution C: 1 g of malonic acid (propanedioic acid) in 10 cm^3 of deionised water.

Ferroin: 0.025 mol dm^{-3} solution.

Teaching tips

Ferroin is phenanthroline ferrous sulfate. Photoflo can be found at any photography shop. It is a surface-active agent used in developing and printing. This oscillating reaction does not need a magnetic stirrer. Solutions need to be freshly prepared before the lesson. Remind students not to move the syringes between chemicals to avoid contamination.

Background theory

Bromate and bromide react with malonic acid to produce bromomalonate. Bromate also reacts with red ferrous dyes to produce blue ferric dye. Bromomalonate and blue ferric dye react to form bromide. Bromide inhibits the reaction of red ferrous dye to blue dye and a red colour is produced. Further details can be found in *New Scientist Guide to Chaos*, p 111. N. Hall (Ed), London: Penguin, 1991.

Safety

Wear eye protection.

Be very careful to insist on eye protection, as long as anybody in class is using a syringe.

Answers

1. Ferroin acts as the dye and gives the solution its colour.
2. Surfactants or surface-active agents usually lower the surface tension of a liquid.

RS•C

An oscillating reaction

Introduction

Several solutions are mixed in a petri dish. After about 5 min, the colour of the solution oscillates between red and blue.

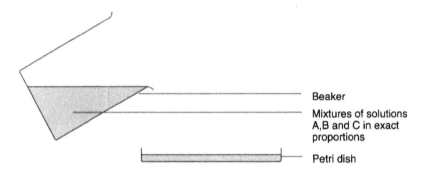

What to do

1. Prepare the oscillating solution as follows: Place 6 cm^3 of solution A in a small beaker using a measuring cylinder.
2. Add 0.5 cm^3 of solution B using a syringe.
3. Add 1.0 cm^3 of solution C using a syringe. A brown colour appears. When it disappears, add 1.0 cm^3 of ferroin using a syringe.
4. Add 1 drop of Photoflo using the dropper pipette.
5. Add enough of this solution in the beaker to a petri dish to half-fill it.
6. Wait for oscillations to begin.

Safety

Wear eye protection. Solutions may be irritant.

Questions

1. What is the role of ferroin in this reaction?
2. What is a surfactant or surface-active agent?

57. Chocolate and egg

Topic

Physical and chemical changes.

Timing

30 min.

Description

Students heat egg white and chocolate.

Apparatus and equipment (per group)

- Two boiling tubes
- 250 cm^3 Beaker
- Bunsen burner
- Tripod
- Gauze
- Test-tube rack
- Test-tube holder.

Chemicals (per group)

- Chocolate
- Egg albumen.

Teaching tips

Egg white can be messy but reacts quickly (~5 min). Chocolate bar squares melt very slowly (>10 min) as does Cadburys Flake. Chocolate drops are faster, grated cooking chocolate is best.

Background theory

This is a good introduction on how to heat safely using a Bunsen burner. It is also a good introduction to physical and chemical change.

Safety

Wear eye protection. Strictly enforce a no tasting rule.

Answers

1. Melting of chocolate.
2. Boiling of egg.
3. Chemical: iron rusting, wood burning *etc*. Physical: ice melting, water-boiling *etc*.

RS•C

Chocolate and egg

Introduction

This experiment shows some changes that happen when different substances are heated.

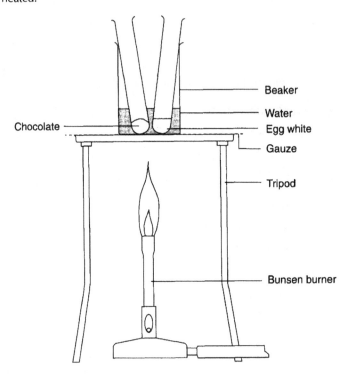

What to record

Make a table of your observations.

Substance	What happens on heating	What happens on cooling

What to do

1. Set up the equipment as shown in the diagram.
2. Heat the boiling tubes in the beaker of water. The water in the beaker should boil.
3. Watch what happens to the substances in the tubes as they are heated.
4. Turn off the Bunsen burner and use the test-tube holder to transfer the tubes to the rack to cool.
5. Watch what happens to the substances in the tubes as they cool.

Safety

Wear eye protection. Do not taste foods in a laboratory. The food or the apparatus may be contaminated.

Questions

1. Which of these changes is reversible? This is called a physical change.
2. Which of these changes produces a new substance? This is called a chemical change.
3. Give one more example of a physical and a chemical change.

58. Catalysis

Topic

Rates of reaction, catalysts.

Timing

45 min

Description

Students compare various catalysts.

Apparatus and equipment (per group)

- ▼ Stopclock (or timer)
- ▼ Dropping pipette. Use the type of teat pipette (usually fitted to Universal Indicator bottles) that does not allow squirting – *eg* Griffin.
- ▼ One 100 cm^3 and one 50 cm^3 measuring cylinder.

Chemicals (per group)

- ▼ Copper(II) sulfate   0.1 mol dm^{-3}
- ▼ Nickel(II) sulfate   0.1 mol dm^{-3}
- ▼ Cobalt(VI) chloride   0.1 mol dm^{-3}
- ▼ Iron(II) sulfate   0.1 mol dm^{-3}
- ▼ Iron(III) nitrate solution 0.1 mol dm^{-3}
- ▼ Sodium thiosulfate   0.1 mol dm^{-3}.

Teaching tips

Ensure the concentrations of solutions are correct and that only one drop of catalyst is used. Only the iron based catalyst is effective. High concentrations or too much catalyst causes the reaction to complete instantaneously. It is best to demonstrate the reaction first.

A light sensor set up as a colorimeter can be used to monitor colour change on the computer. The result, in the form of graphs on the computer, provides very useful material for analysis using software. A light sensor clamped against a plastic cuvette filled with the reactants substitutes for a colorimeter. The data logging software should show the colour change occurring on a graph and this tends to yield more detail than the standard end-point approach. The rate of change can be measured from the graph slope or the time taken for a change to occur.

Background theory

Concept of a catalyst.

Safety

Wear eye protection.

Answers

1. Copper(II) sulfate solution.
2. The colour of the solution may obscure the end point, the reaction may become too fast to compare the catalysts.
3. Lower temperature, lower concentration, smaller amount of catalyst.

Catalysis

Introduction

In this experiment the speed of a reaction is measured. Various metals in solution are tested as possible catalysts.

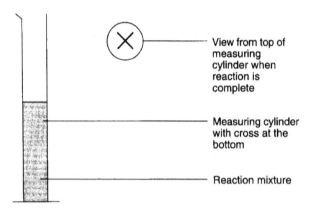

What to record

Complete the following table.

	No catalyst	Nickel(II) sulfate	Copper(II) sulfate	Iron(II) sulfate	Cobalt(II) chloride
Time (s)					

What to do

1. Draw a cross on a piece of scrap paper and put it underneath the measuring cylinder so it can be seen when looking down the cylinder from the top.
2. Using a 50 cm^3 measuring cylinder measure 50 cm^3 of iron(III) nitrate solution.
3. Using a 100 cm^3 measuring cylinder measure 50 cm^3 of sodium thiosulfate solution.
4. Pour the iron(III) nitrate solution into the sodium thiosulfate solution and start the timer.
5. Look through the reaction mixture from above until the cross can first be seen. Stop the timer and record the time.
6. Repeat this experiment but add one drop of catalyst to the iron(III) nitrate solution before mixing. Test the various catalysts and fill in the table.

Safety

Wear eye protection.

Questions

1. Which metal compound is the best catalyst?
2. Why were only very dilute solutions of metal compounds used?
3. A catalyst speeds up a reaction. Suggest one way of slowing down this reaction.

RS•C

59. A Cartesian diver

Topic

Particulate nature of matter, liquids and gases.

Timing

Variable.

Description

Students make a Cartesian diver from a fizzy drink bottle and a plastic pipette. This experiment illustrates how gases are more compressible than liquids.

Apparatus and equipment (per group)

- ▼ 5 cm^3 Plastic disposable pipette (available from Hogg)
- ▼ Hex nut (from most DIY superstores 11mm across face to face)
- ▼ One 2 dm^3 or 1.5 dm^3 or 1 dm^3 soft drink bottle with lid (clear plastic)
- ▼ 250 cm^3 Plastic beaker
- ▼ Access to scissors.

Teaching tips

It is important to adjust the diver so that it barely floats in the beaker. If the diver requires a strong squeeze to make it sink there is not enough water in the diver. If the diver sinks then it has too much water inside. The diver should be about half full with water.

Background theory

The diver sinks if it is more dense than the surrounding water and rises if it is less dense.

Safety

Mop up any water spillage from the floor.

Answers

1. Air is compressed and the volume of air is reduced.
2. When compared with those in liquids, the gas particles which make up air are a large distance apart. It is therefore easier to squash them closer together thus reducing the volume.
3. When the bottle is squeezed pressure in the water pushes on the pocket of air inside the diver. The volume of air is reduced and this allows more water into the diver. This makes the diver more dense and it therefore sinks. When the pressure is released, the air expands taking up a larger volume. Water is pushed out of the diver which becomes less dense and therefore floats in the water.

A Cartesian diver

Introduction

This is an experiment named after René Descartes (1596–1650). Descartes was a French scientist and philosopher. The Cartesian diver can be used to illustrate the behaviour of gases and liquids when compressed. In this experiment a Cartesian diver is constructed and some of the properties observed.

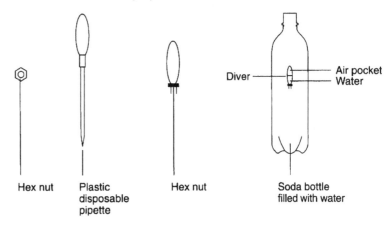

What to do

1. Screw the hex nut onto the base of the pipette until it is held tightly in place.
2. Cut off all but 1 cm of the pipette stem. (This is the diver.)
3. Place the diver in a beaker of water. Squeeze the bulb of the pipette to force air out and release to allow water up into the diver. Repeat this until the diver is about half full of water.
4. Does the diver still float? If adjusted properly the diver should barely float in the water. If it sinks squeeze a little water out.
5. Carefully transfer the diver to the soda bottle that is full to the brim with water. Take care not to lose water from the diver. Place the cap on the bottle.
6. Use both hands and squeeze the bottle. Watch the diver sink when the bottle is squeezed, or float when pressure is released.

Safety

Wipe up any water spillage.

Questions

1. What happens to the air in the diver when the bottle is squeezed?
2. Why does the air behave in this way?
3. Write a sentence that explains how the Cartesian diver works.

RS•C

60. Neutralisation of indigestion tablets

Topic

Neutralisation.

Timing

30 min.

Description

Students neutralise an indigestion tablet with hydrochloric acid.

Apparatus and equipment (per group)

- ▼ Pestle and mortar (small)
- ▼ 250 cm^3 Conical flask
- ▼ Burette
- ▼ Burette stand.

Chemicals (per group)

- ▼ One indigestion tablet
- ▼ Methyl orange indicator
- ▼ Hydrochloric acid 0.5 mol dm^{-3}.

Teaching tips

This experiment could be adapted for an investigation by using various brands of indigestion tablet. Students could then find which is most effective. One suggestion is to calculate how much active ingredient a tablet contains. A cost benefit analysis may be attempted if the cost of each tablet is calculated. If burettes are used with students aged 11–14 it is advisable to provide them ready clamped and filled to the zero mark.

Background theory

Indigestion tablets are alkali in solution, they neutralise excess stomach acid.

Safety

Wear eye protection.

# Neutralisation of indigestion tablets

Introduction

Indigestion is caused by excess acid in the stomach. The tablets neutralise some of this acid. In this experiment the amount of acid neutralised by one tablet is found. This may be considered a direct measurement of the effectiveness of the tablet.

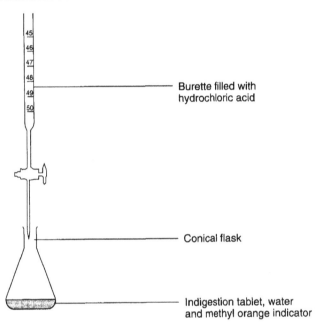

What to record

Record the volume and concentration of the acid added.

What to do

1. Crush a tablet using a mortar and pestle and carefully transfer it to a conical flask.
2. Add about 25 cm^3 of water and three drops of methyl orange indicator.
3. While slowly swirling the flask, add acid from the burette, 0.5 cm^3 at a time.
4. Continue adding acid in 0.5 cm^3 portions until the liquid goes red and stays red for one minute.
5. Record the volume of acid used.

Safety

Wear eye protection.

Question

1. Samples of various brands of indigestion tablets and the cost of each packet are provided. Describe how this experiment could be used to determine which brand represents the best value.

RS•C

61. Mass conservation

Topic

Mass conservation.

Timing

20 min.

Description

Students react potassium iodide with lead nitrate. They weigh reactants and products to show there is no mass change.

Apparatus and equipment (per group)

- ▼ Two < 100 cm^3 beakers
- ▼ 10 cm^3 Measuring cylinder
- ▼ Access to an accurate balance (0.01g).

Chemicals (per group)

- ▼ Potassium iodide solution   0.01 mol dm^{-3}
- ▼ Lead nitrate solution   0.009 mol dm^{-3} (**Harmful**).

Teaching tips

Students get confused with errors in an experiment. This is a good opportunity to reinforce the idea of errors and that small differences do not necessarily indicate weight loss or gain. The teacher can state that if weight is gained or lost, the results would be repeatable, and therefore the same for all groups in the class.

Emphasise the process of not trying to weigh exact amounts of reactants, you simply require an accurate mass.

Background theory

Accuracy in measurements, errors, and repeatability.

Safety

Wear eye protection. Ensure students wash their hands thoroughly after the experiment.

Answers

1. Yes, this is confirmed by the colour change.
2. Potassium iodide + lead nitrate → potassium nitrate + lead iodide
3. Although at some degree of accuracy the experiment appears to show a mass change this is within the limits of experimental error and these results tend to indicate no mass change.

Mass conservation

Introduction

The aim of this experiment is to show that mass is not gained or lost in a chemical reaction.

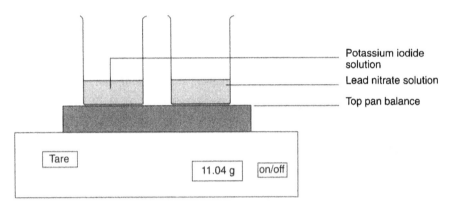

What to record

Record the total mass of the reactants and the products.

What to do

1. Measure approximately 5 cm^3 of potassium iodide in the measuring cylinder and pour into one beaker.
2. Rinse the measuring cylinder.
3. Measure approximately 5 cm^3 of lead nitrate in the measuring cylinder and pour into another beaker.
4. Zero the balance then place both beakers on the balance at the same time. Accurately measure their total mass.
5. Take them off the balance.
6. Carefully pour the contents of one beaker into the other making sure there is none spilt.
7. Zero the balance again, place both beakers back on the pan and measure the mass.

Safety

Wear eye protection. Solutions of lead compounds may be toxic – wash your hands thoroughly at the end of the lesson.

Questions

1. Has a chemical reaction occurred?
2. Complete the word equation:
   potassium iodide + lead nitrate → _____ + _____
3. Comment on your result.

RS•C

62. Metals and acids

Topic

Metals, salts.

Timing

60 min.

Description

Zinc and sulfuric acid are reacted to form zinc sulfate.

Apparatus and equipment (per group)

Lesson 1
- 100 cm^3 Conical Flask
- 250 cm^3 Beaker
- Labels
- 50 cm^3 or 100 cm^3 Measuring cylinder
- Filter funnel & paper
- Evaporating basin
- Crystallising dish
- Bunsen burner
- Tripod
- Gauze.

Lesson 2
- Hand lens.

Chemicals (per group)

- Sulfuric acid 1 mol dm^{-3} (**Irritant**)
- Eight lumps of zinc.

Teaching tips

Immerse the zinc lumps in copper(II) sulfate solution (very dilute) prior to the lesson for half an hour or so. They will react much more readily.

Background theory

Acid + metal → salt + water

Safety

Wear eye protection. Care with hot acid.

Answers

1. Zinc + sulfuric acid → zinc sulfate + hydrogen
2. (a) Zinc + hydrochloric acid → zinc chloride + hydrogen
   (b) Magnesium + sulfuric acid → magnesium sulfate + hydrogen
3. $Zn + H_2SO_4 \rightarrow ZnSO_4 + H_2$
   $Zn + 2HCl \rightarrow ZnCl_2 + H_2$
   $Mg + H_2SO_4 \rightarrow MgSO_4 + H_2$

Metals and acids

Introduction

Many, but not all, metals react with acids. Hydrogen gas is formed and the metal reacts with the acid to form a salt.

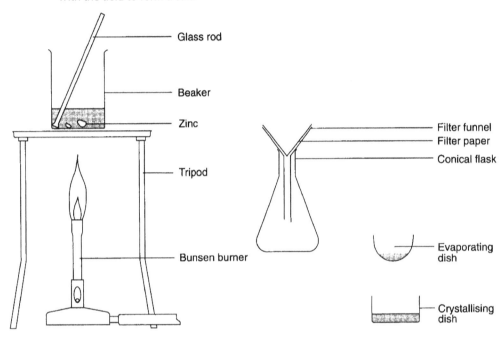

What to do

Lesson 1
1. Measure 50 cm^3 of dilute sulfuric acid and pour it into the beaker. Warm this acid but turn off the Bunsen burner before it reaches the boiling point.
2. Carefully remove the beaker of acid from the tripod and stand it on the bench.
3. To this acid, add two lumps of zinc.
4. If all the zinc reacts, add two more lumps. Add more zinc until no more bubbles form. The acid is now used up.
5. Filter into the conical flask to remove the excess zinc and transfer the filtrate into an evaporating basin.
6. Gently heat the filtrate. Dip in a glass rod and hold it up to cool. When small crystals form on the glass rod stop heating.
7. Pour the solution into a crystallising dish. Label the dish and leave it to crystallise for next lesson.

Lesson 2
1. Examine the crystals with a hand lens.

Safety

Wear eye protection. Care with hot acid.

Questions

1. Write a word equation for the reaction between zinc and sulfuric acid.
2. Write word equations for the reactions of:
   (a) zinc and hydrochloric acid.
   (b) magnesium and sulfuric acid.
3. Write equations for these three reactions using chemical formulas.

63. Solid mixtures – a lead and tin solder

Topic

Mixtures, alloys and their properties.

Timing

30 min.

Description

The melting points of lead, tin and solder (an alloy of the two) are compared.

Apparatus and equipment (per group)

- ▼ Tripod
- ▼ Pipe clay triangle
- ▼ Crucible lid
- ▼ Bunsen burner.

Chemicals (per group)

- ▼ Small piece of tin
- ▼ Small piece of lead
- ▼ Small piece of solder (use flux free solder).

Teaching tips

One problem which was identified in the laboratory is that students forget the identity of each piece of metal. It is important to remind students to remember which is which. Solder must not contain flux.

Background theory

Metal alloys are solid solutions, they are homogenous mixtures normally mixed in the liquid state.

Safety

Wear eye protection. Ensure good ventilation. Warn your class to be careful with drops of molten metal. Warn asthmatics, who should preferably use a fume cupboard.

Answers

1. Solder.
2. Lead.
3. Solder.
4. Lead.
5. Lowest melting point is solder, then tin (231.9 °C), then lead (327.4 °C).

Solid mixtures – a lead and tin solder

Introduction

Electrical solder is an alloy of tin and lead. In this experiment a simple method is used to compare the melting points of lead, tin and solder.

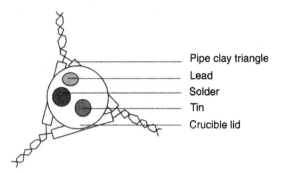

What to record

▼ What was done.
▼ The order in which they melt.
▼ The order in which they solidify.

What to do

1. Place a small piece of lead (Pb), tin (Sn), and solder (a Pb/Sn mixture) on an inverted crucible lid over a Bunsen flame. Remember which is which.
2. Observe the three metals, to see which one melts first.
3. When all three are molten, turn off the Bunsen burner and allow the metals to cool.
4. Observe the order in which they solidify.

Safety

Wear eye protection.

Questions

1. Which of the three metals melts first?
2. Which of the three metals melts last?
3. Which of the three metals solidifies last?
4. Which of the three metals solidifies first?
5. Write down the metals in order of their melting points, lowest melting point first.

64. The effect of temperature on reaction rate

Topic

Rate of reaction.

Timing

60 min.

Description

Sodium thiosulfate solution is reacted with acid, a precipitate of sulfur forms. The time taken for a certain amount of sulfur to form is used to indicate the rate of the reaction. The effect of temperature on the rate of reaction can be investigated.

Apparatus and equipment (per group)

- 250 cm^3 Conical flask
- 10 cm^3 Measuring cylinder
- 50 cm^3 Measuring cylinder.

Chemicals (per group)

- Sodium thiosulfate solution   40 g dm^{-3}
- Hydrochloric acid   2 mol dm^{-3} (**Irritant**).

Teaching tips

The method for this experiment is best understood when the teacher demonstrates it first. The end-point can be measured with a light sensor connected to a data-logger.

A light sensor set up as a colorimeter can be used to monitor the precipitation on a computer – clamp a light sensor against a plastic cuvette filled with the reactants. The result, in the form of graphs on the computer, provides very useful material for analysis using data logging software. The software shows the change on a graph and this tends to yield more detail than the end-point approach used in this experiment. The rate of change can be measured from the graph slope or the time taken for a change to occur.

Background theory

Basic collision theory.

Safety

Wear eye protection. Sulfur dioxide (**Toxic gas**) forms as a by-product. Ensure good ventilation. Warn asthmatics, who should preferably use a fume cupboard.

As soon as the reaction is complete pour the solutions away, preferably into the fume cupboard sink. Wash away with plenty of water. This is particularly important with solutions used at higher temperatures.

The effect of temperature on reaction rate

Introduction

In this experiment the effect of temperature on the rate of reaction between sodium thiosulfate and hydrochloric acid is investigated.

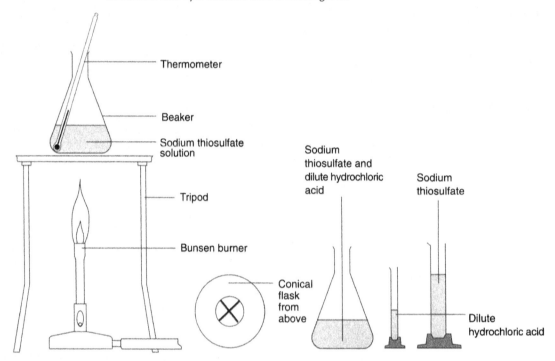

What to record

Record your results in the table.

Initial temperature of the mixture in the flask/°C	Final temperature of the mixture in the flask/°C	Average temperature of the mixture in the flask/°C	Time taken for the cross to disappear/s	1/time taken /s^{-1}

What to do

1. Put 10 cm^3 of sodium thiosulfate solution and 40 cm^3 of water into a conical flask. Measure 5 cm^3 of dilute hydrochloric acid in a small measuring cylinder.
2. Warm the thiosulfate solution in the flask if necessary to bring it to the required temperature. The object is to repeat the experiment five times with temperatures in the range 15–65 °C.
3. Put the conical flask over a piece of paper with a cross drawn on it.

4. Add the acid and start the clock. Swirl the flask to mix the solutions and place it on a piece of white paper marked with a cross. Take the initial temperature of the mixture.
5. Look down at the cross from above. When the cross disappears, stop the clock and note the time taken. Record the final temperature of the mixture in the flask.
6. As soon as possible, pour the solution down the sink (in the fume cupboard if possible) and wash away.

Safety

Wear eye protection. Take care not to inhale fumes.

Questions

1. For each set of results, calculate the value of 1/time. (This value can be taken as a measure of the rate of reaction for this experiment).
2. Plot a graph of 1/time on the vertical (y) axis and average temperature on the horizontal (x) axis.

RS•C

65. The effect of concentration on reaction rate

Topic

Rate of reaction.

Timing

60 min.

Description

Sodium thiosulfate solution is reacted with acid – a sulfur precipitate forms. The time taken for a certain amount of sulfur to form can be used to indicate the rate of the reaction.

Apparatus and equipment (per group)

- ▼ 250 cm^3 Conical flask
- ▼ 100 cm^3 Measuring cylinder.

Chemicals (per group)

- ▼ Sodium thiosulfate solution — 50 g dm^{-3}
- ▼ Hydrochloric acid — 2 mol dm^{-3}.

Teaching tips

The method for this experiment is best understood when the teacher demonstrates it first.

A light sensor can be used to monitor the precipitation on a computer. The result, in the form of graphs on the computer, can be analysed using data logging software. A light sensor clamped against a plastic cuvette filled with the reactants substitutes for a colorimeter. The data logging software shows the turbidity on a graph and this tends to yield more detail than the standard end-point approach. The rate of change can be measured using the slope of the graph or the time taken for a change to occur.

Background theory

Basic collision theory.

Safety

Wear eye protection. Sulfur dioxide (**Toxic gas**) forms as a by-product. Ensure good ventilation. Warn asthmatics, who should preferably use a fume cupboard.

As soon as the reaction is complete pour the solutions away, preferably into the fume cupboard sink. Wash away with plenty of water.

The effect of concentration on a reaction rate

Introduction

In this experiment, the effect of the concentration of sodium thiosulfate on the rate of reaction is investigated.

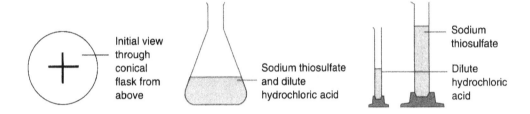

What to record

1. Complete the table:

Volume of sodium thiosulfate solution/cm^3	Volume of water/cm^3	Time taken for cross to disappear /s	Original concentration of sodium thiosulfate solution/g dm^{-3}	1/time taken /s^{-1}
50	0		50	
40	10			
30	20			
20	30			
10	40			

What to do

1. Put 50 cm^3 of sodium thiosulfate solution in a flask.
2. Measure 5 cm^3 of dilute hydrochloric acid in a small measuring cylinder.
3. Add the acid to the flask and immediately start the clock. Swirl the flask to mix the solutions and place it on a piece of paper marked with a cross.
4. Look down at the cross from above. When the cross disappears stop the clock and note the time. Record this in the table.
5. Repeat this using different concentrations of sodium thiosulfate solution. Make up 50 cm^3 of each solution. Mix different volumes of the sodium thiosulfate solution with water as shown in the table.
6. As soon as possible, pour the solution down the sink (in the fume cupboard if possible) and wash away.

Safety

Wear eye protection. Take care not to inhale fumes.

Questions

1. Calculate the concentration of sodium thiosulfate in the flask at the start of each experiment. Record the results in the table.
2. For each set of results, calculate the value of 1/time. (This value can be taken as a measure of the rate of reaction).
3. Plot a graph of 1/time taken on the vertical (y) axis and concentration on the horizontal (x) axis.

66. The effect of heat on metal carbonates

Topic

Reactions of carbonates.

Timing

40 min.

Description

This experiment involves a comparison between carbonates of reactive metals such as sodium and potassium with the carbonates of less reactive metals such as lead and copper.

Apparatus and equipment (per group)

- ▼ Two test-tubes
- ▼ Delivery tube (right angled)
- ▼ Bunsen burner.

Chemicals (per group)

- ▼ Sodium carbonate
- ▼ Potassium carbonate
- ▼ Lead carbonate (**Toxic**)
- ▼ Copper carbonate (**Harmful**).

Teaching tips

Remind students to clamp the test-tube well away from where the tube is heated. The time taken for the limewater to go milky is a measure of the ease of decomposition.

Background theory

Basic reaction pattern:

Metal carbonate → metal oxide + carbon dioxide

Reactive metals form strong bonds that are difficult to break.

Safety

Wear eye protection. Remind students that the liquid may 'suck-back'. Avoid raising the dust of lead compounds. Wash hands thoroughly at the end of the experiment.

Answers

1.

Carbonate tested	Colour of metal carbonate before heating	Gas evolved, if any	Decomposition easy or difficult	Colour of solid after heating
Sodium carbonate	White	None	Difficult	White
Lead carbonate	White	Carbon dioxide	Easy	Yellow
Potassium carbonate	White	None	Difficult	White
Copper carbonate	Bluish-green	Carbon dioxide	Easy	Black

2. Air in the tube expands and bubbles through the limewater.
3. To prevent 'suck back'.

The effect of heat on metal carbonates

Introduction

Metal carbonates decompose when heated. Some carbonates are more reactive than others. The aim of this experiment is to compare the reactivity of some different metal carbonates.

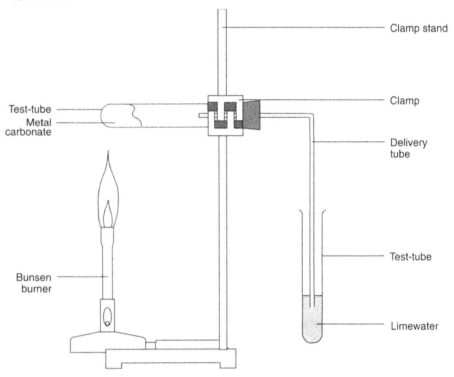

What to record

Complete the table

Carbonate tested	Colour of metal carbonate before heating	Gas evolved, if any	Decomposition easy or difficult	Colour of solid after heating
Sodium carbonate				
Lead carbonate				
Potassium carbonate				
Copper carbonate				

RS•C

What to do

1. Put a large spatula of the carbonate to be tested in a test-tube.
2. Fit a delivery tube and then clamp the tube so that the delivery tube dips into a second test-tube containing limewater.
3. Heat the solid gently at first then more strongly.

Safety

Wear eye protection.

Remove the delivery tube by lifting the clamp stand as soon as heating is stopped. Some metal compounds are toxic. Avoid raising dust. Wash hands thoroughly at the end of the experiment.

Questions

1. Why do some gas bubbles pass through limewater when heating is first started?
2. Why must the delivery tube be removed as soon as heating is stopped?

67. The change in mass when magnesium burns

Topic

Oxidation.

Timing

45 min.

Description

Students react magnesium with oxygen. They weigh the magnesium and the magnesium oxide product.

Apparatus and equipment (per group)

- Crucible and lid
- Access to a balance
- Tongs
- Pipe clay triangle
- Tripod
- Bunsen burner.

Chemicals (per group)

- Magnesium ribbon 10–20 cm.

Teaching tips

Demonstrate the technique of lifting the lid a little and explain the purpose of the lid is to minimise any loss of smoke. The lid is lifted to allow access for the oxygen.

Background theory

Magnesium combines with oxygen from the air. There is an increase in mass. The increase is the mass of oxygen that combines with magnesium.

Safety

Wear eye protection.

Answers

1. To remove any magnesium oxide.
2. To reduce the loss of magnesium oxide smoke particles.
3. To ensure enough oxygen is supplied to the reacting magnesium.
4. When reaction appears complete, weigh the crucible, then reheat and weigh again. If the reaction is complete there is no increase in weight.

Extension

This experiment can be used to determine the formula of magnesium oxide. See experiment number 90.

The change in mass when magnesium burns

Introduction

Many areas of chemistry involve careful measurement. One example is measuring the change in mass before and after a chemical reaction. This experiment shows how the mass of magnesium changes when it combines with oxygen.

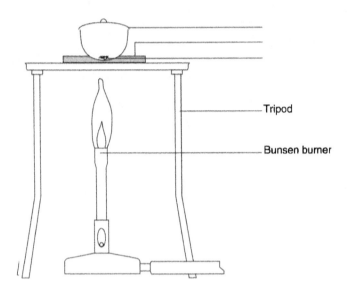

What to record

What was done.

The mass of the magnesium and the magnesium oxide product.

What to do

1. Clean a 10–20 cm length of magnesium ribbon with emery cloth to remove the oxide layer. Loosely coil it.
2. Weigh a clean crucible and lid. Place the magnesium inside and reweigh.
3. Heat the crucible for 5–10 minutes, lifting the lid a little from time to time with tongs. Ensure that as little product as possible escapes.
4. Continue heating until glowing ceases.
5. Cool the crucible and reweigh.

Safety

Wear eye protection.

Questions

1. Why is the magnesium ribbon cleaned before the experiment?
2. Why is the lid needed?
3. Why is the lid lifted from time to time?
4. How could weighing be used to show when the reaction is complete?

68. The volume of 1 mole of hydrogen gas

Topic

Molar volume.

Timing

40 min.

Description

Students react magnesium quantitatively with hydrochloric acid. They collect the hydrogen and calculate the molar volume.

Apparatus and equipment (per group)
- Burette
- Burette stand
- Water bath.

Chemicals (per group)
- Hydrochloric acid   2 mol dm^{-3} (**Irritant**)
- Magnesium ribbon   0.02–0.04 g (~3.5mm standard ribbon).

Teaching tips

You should demonstrate the procedure beforehand. The inversion is not difficult. Rest the end of the burette on the lip of the beaker and swing the tap end round and upward to a vertical position. It is important that the liquid level starts on the graduated scale of the burette. If the liquid level is not on the scale, opening the tap momentarily allows the liquid to drop onto the scale.

Background theory

Volume of one mole of gas at standard temperature and pressure, stp, (0 °C, 101,500 N m^{-2}) is 22.4 dm^3.

At room temperature and average pressure, rtp, the students can expect an answer of approximately 24 dm^3.

Students that are more able may be able to use the equation

$P_1V_1/T_1 = P_2V_2/T_2$ to find the volume at stp. The temperature and pressure in the laboratory need to be measured.

Safety

Wear eye protection.

Answers

1. Expect rtp molar volume to be approximately 24 dm^3.

RS•C

The volume of 1 mole of hydrogen gas

Introduction

One mole of any gas occupies the same volume when measured under the same conditions of temperature and pressure. In this experiment, the volume of one mole of hydrogen is calculated at room temperature and pressure.

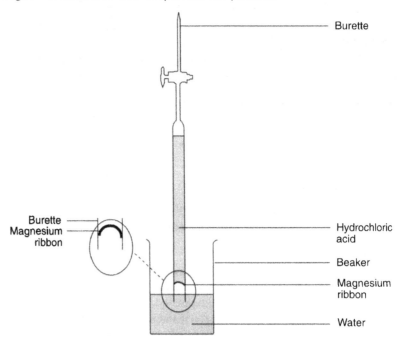

What to record

What was done.

The mass of magnesium used and the volume of hydrogen produced.

What to do

1. Clean a piece of magnesium ribbon about 3.5 cm long and weigh accurately. (This should weigh between 0.02 and 0.04 g; if not adjust the amount used.)
2. Measure 25 cm^3 of dilute hydrochloric acid (**Irritant**) into the burette. Carefully add 25 cm^3 of water on top of this.
3. Push the magnesium into the end of the burette so it will stay in position with its own tension.
4. Add 50 cm^3 of water to a 250 cm^3 beaker.
5. Quickly invert the burette into the water. If this is done quickly and carefully very little is lost. It is important that the liquid level in the burette starts on the graduated scale. If it is not on the scale; momentarily open the tap, this allows the level to drop). Clamp the burette vertically.
6. Take the burette reading (care: it is upside down!)
7. Observe the magnesium react as the acid diffuses downwards, wait until all the magnesium has reacted.

8. Note the new volume on the burette (care: it is upside down).
9. Record your results.

Safety

Wear eye protection.

Questions

The equation for the reaction is

$Mg + 2HCl \rightarrow MgCl_2 + H_2$

The relative atomic mass of magnesium is 24.

1. Copy out and fill in the gaps:
   ____ g Magnesium has produced ____ cm^3 hydrogen
   ____ /24 moles magnesium produces ____ cm^3 hydrogen.
   1 mole magnesium produces _____ cm^3 hydrogen which is the volume of one mole of hydrogen gas.

RS•C

69. How much air is used up during rusting?

Topic

Oxidation, rusting, percentage oxygen in the air.

Timing

Two lessons.

Description

Students set up an experiment to find the percentage of air that is used during rusting.

Apparatus and equipment (per group)

- ▼ Test-tube
- ▼ 100 cm^3 Beaker
- ▼ Ruler.

Chemicals (per group)

- ▼ Iron wool.

Teaching tips

This experiment can be used to find the percentage of oxygen in air. The test-tube starts full of air. After about one week students measure the new length of air. Care is needed to avoid lifting the test-tube out of the water. Test-tubes get rust stains. These can be cleaned with a 'Stain Devil'®.

Background theory

Students should understand that rusting is an oxidation of iron to form iron oxide.

Answers

1. Iron + oxygen → iron oxide
2. Approximately 20 per cent.
3. Leave the experiment for a further period. Show that no more air has reacted.

How much air is used during rusting

Introduction

This experiment illustrates how much of the air is used in the rusting process. It is the oxygen component of air which reacts in the rusting process. This experiment allows calculation of the percentage of oxygen in the air.

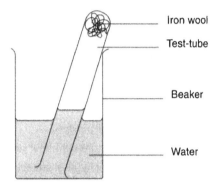

What to record

Initial length of column of air	
Final length of column of air in the tube	

What to do

1. Place approximately 3 cm depth of iron wool in the bottom of a test-tube. Wet the iron wool with water.
2. Invert the test-tube in a beaker of water (approximately 20 cm^3).
3. Measure the length of the column of air.
4. Leave the test-tube for at least one week, and then measure the new length of the column of air. Take care not to lift the test-tube out of the water.

Questions

1. Write a word equation for this reaction.
2. Calculate the percentage of oxygen in air.
3. How could it be shown that the reaction is complete?

RS•C

70. Making a photographic print

Topic

Halogens, alkali metals.

Timing

40 min.

Description

Students produce a photographic print using silver chloride.

Apparatus and equipment (per group)

- ▼ Paper
- ▼ Two small brushes
- ▼ Hair drier (do not allow students to bring these from home)
- ▼ Ultraviolet light (or sunlight) (ultraviolet lamps must be screened so that students cannot look directly at the source)
- ▼ Large glass sheet.

Chemicals (per group)

- ▼ Silver nitrate solution (the silver nitrate dilution is 1.3 g in 10 cm^3 of deionised water). (**Corrosive**)
- ▼ Potassium chloride solution (the potassium chloride dilution is 0.5 g in 10 cm^3 of deoinised water).

10 cm^3 of each solution is enough for about forty prints of 5 cm x 5 cm.

Teaching tips

A good quality paper such as watercolour paper must be used.

Any flat objects are ideal to place on the paper, including:

Paper shapes – students can design their own; and

Dried leaves or pressed dried flowers.

A clean sheet of glass can be used to cover the prints during exposure to the light; this prevents the paper curling during exposure.

It is advisable to use a photographic fixer once the exposure is complete. The teacher (or a technician) should do this. Acufix is suitable and available from any photographic supplier. A 1:7 dilution with water at room temperature is adequate and enables the students to keep their prints and place them in their laboratory books.

Background theory

Silver chloride, silver bromide and silver iodide (silver halides) are reduced to silver by the action of light, X-rays and radiation from radioactive substances. They are used to make photographic film and photographic paper.

A solution of potassium chloride is put onto the paper and dried. A solution of silver nitrate is then used in the same way. The two compounds undergo a precipitation reaction, giving a slight creamy discolouration to the paper.

$KCl + AgNO_3 \rightarrow AgCl + KNO_3$

The paper is then dried again. When the paper is exposed to light, X-rays or radioactive substances, the silver chloride decomposes and the silver metal colours the paper. The areas of the paper that have been covered are not exposed to the energy source and remain unchanged. If a photographic negative is used, then the coloration occurs to varying degrees.

Safety

Wear eye protection. Gloves should be worn to handle the silver nitrate and the coated paper. Silver nitrate stains your hands.

Answers

1. The paper goes slightly creamy in colour.
2. Potassium chloride + silver nitrate → silver chloride + potassium nitrate
3. When the paper is exposed to any of the energy sources mentioned, the silver chloride decomposes and the silver metal colours the paper.

RS•C

Making a photographic print

Introduction

Only a very small amount of energy is needed to break down silver halide compounds (silver chloride, silver bromide or silver iodide) to the silver metal. This small amount of energy is available from many sources including light, X-rays and radiation from a radioactive substance. The above three silver halides can be used to make photographic film and photographic paper. In this experiment, a photographic print is produced.

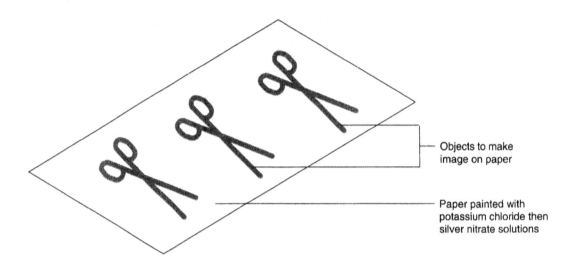

— Objects to make image on paper

— Paper painted with potassium chloride then silver nitrate solutions

What to do

1. In a darkened room, take a piece of paper and paint one side of it with the potassium chloride solution. Dry the paper with a hairdryer.
2. Paint the dried paper, on the same side, with the silver nitrate solution. Dry the paper with a hairdryer.
3. Place your chosen objects(s) on top of the paper and place under an ultraviolet light for 30 min or in sunlight for 2 h.
4. Switch off the ultraviolet light (if used) and remove the objects from the top of the paper. Observe what has happened.

Safety

Wear eye protection and protective gloves. Care with ultraviolet light; do not look directly at the light, it can damage your eyes.

Questions

1. What happens to the paper when the silver nitrate is painted onto it?
2. Write a word equation for the above reaction.
3. Explain what happens when the paper is exposed to light.

71. 'Smarties' chromatography

Topic

Separation.

Timing

30 min.

Description

Students separate the dyes in Smarties food colouring using chromatography paper with water solvent.

Apparatus and equipment (per group)

- ▼ 'Smarties'
- ▼ Paint brush
- ▼ 250 cm^3 Beaker
- ▼ Two paper clips
- ▼ Chromatography paper (approximately 200 mm x 100 mm).

Teaching tips

Take care to avoid smudging; small intense spots are best. The paper must be labelled in pencil.

Small bottles of liquid food colouring can be purchased from supermarkets. These are water soluble and can be used as an alternative to Smarties.

Slotted chromatography paper (Whatman) is the best for this experiment.

Background theory

Students should have a basic understanding of chromatography theory. This experiment can be a useful introduction to this separation method.

Safety

No eating in the laboratory.

Answers

1. Some dyes are mixtures and separate on the paper, other dyes are single substances.
2. Some dyes are more soluble in water; some dyes adhere to the paper more strongly.
3. It is possible to identify all the dyes using a list of E numbers. Smarties dyes are regularly changed. A table of E numbers for dyes could be used, also access to the Smarties tube or packet.

RS•C

'Smarties' chromatography

Introduction

In this experiment dye is removed from the surface of various Smarties. A spot of each colour is put on a piece of chromatography paper and water is allowed to soak up the paper. The results show which mixtures are used to produce particular colours for the Smarties.

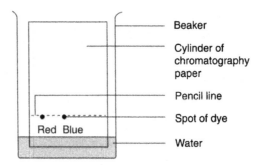

What to record

Record the dyes used to make each colour.

What to do

1. Draw a pencil line 1 cm from the bottom of the chromatography paper.
2. Use a clean paintbrush and clean water to remove the colour from a Smartie. Paint the colour in a small spot on to the line on the chromatography paper.
3. Clean the brush and paint the colour of another Smartie on a small spot about 2 cm from the previous spot. Repeat this until all the colours are on the paper.
4. Using the pencil write the name of each colour by the corresponding spot.
5. Roll the paper into a cylinder, hold in place with paper clips. Put the cylinder in a beaker containing 1 cm of water. Allow the water to rise up the paper.
6. When it reaches the top take the cylinder out of the water, carefully unroll it and examine it.

Safety

Do not eat in the laboratory.

Questions

1. Why do some dyes separate into different colours yet others do not?
2. Why do some dyes move further up the chromatography paper than others?
3. Look on the side of the Smarties packet for the list of coloured dyes used. Try and identify which dyes correspond to the spots on the chromatogram.

72. The decomposition of magnesium silicide

Topic

Oxidation, activation energy, Group 4 elements.

Timing

20 min.

Description

Magnesium silicide is produced and then decomposed using dilute acid. The silane produced ignites spontaneously on contact with air.

Apparatus and equipment (per group)

- 1 dm^3 Beaker
- 100 cm^3 Measuring cylinder
- Ignition tube
- Tongs
- Access to safety screens
- Clamp stand base.

Chemicals (per group)

- Dry magnesium powder (**Highly flammable**)
- Silicon powder
- Hydrochloric acid 2 mol dm^{-3} (**Irritant**).

Teaching tips

Keep windows open, darken the room at the decomposition stage. Observe the exothermic reaction of magnesium with silicon. If it is not possible for all the students to work behind safety screens, then the preparation of the magnesium silicide should be demonstrated. The class can use pre-prepared samples of magnesium silicide.

Background theory

This is an effective experiment to initiate discussion when studying rates of reaction, energy changes, energy diagrams and the concept of activation energy.

Safety

Wear eye protection. SiO$_2$ is produced as a fine dust. When the powders are heated the tube should be tapped down otherwise the powders are ejected from the tube. Care when breaking ignition tube. Teachers should keep careful control over the distribution of magnesium powder. The reaction between magnesium and silicon should be done behind a safety screen.

Answers

1. A spontaneous reaction occurs.
2. The activation energy for CH$_4$ and O$_2$ is higher than for SiH$_4$ and O$_2$.

The decomposition of magnesium silicide

Introduction

This experiment illustrates a reaction with low activation energy. Magnesium reacts with silicon to produce magnesium silicide. This then decomposes in dilute acid to produce silane, which spontaneously combusts on contact with air.

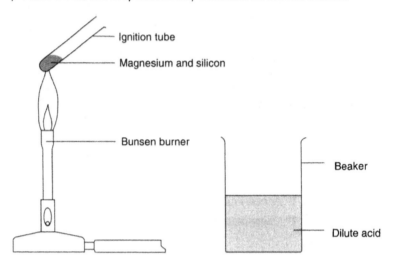

What to do

1. Carefully heat equal amounts of silicon and magnesium powder (1 spatula of each) in an ignition tube (behind a safety screen).
2. Use the clamp stand base to crush the cooled tube between two sheets of paper.
3. Put 100 cm^3 of dilute hydrochloric acid into a 1 dm^3 beaker and add 800 cm^3 of tap water.
4. Put the crushed ignition tube with its contents into the dilute acid, stand back and observe. (Do not inhale the fumes, use a fume cupboard if possible, or ensure the room is well ventilated.)

Safety

Wear eye protection.

Questions

The equations for this reaction are:

$2Mg + Si \rightarrow Mg_2Si$

$Mg_2Si + 2H_2O \rightarrow SiH_4(g) + 2MgO$

$SiH_4 + 2O_2 \rightarrow SiO_2 + 2H_2O$

1. What happens when silane meets oxygen?
2. Why does methane (CH_4) not behave in the same way?

Classic chemistry experiments

RS•C

73. An example of chemiluminescence

Topic

Energy, preparation of solutions, measurement.

Timing

20 min.

Description

Students prepare the solutions in the laboratory. In a darkened room the solutions are mixed to produce chemiluminescence.

Apparatus and equipment (per group)

- Two 10 cm^3 measuring cylinders
- Three 100 cm^3 beakers.

Chemicals (per group)

- Luminol (3 aminobenzene-1,2-dicarboxylic hydrazide)
- Fluorescein
- Methylene blue.
- Sodium hydroxide solution 2 mol dm^{-3}
- Potassium hexacyanoferrate(III) solution 0.1 mol dm^{-3}.

Solution A (**Irritant**): dissolve 0.1 g Luminol in 20 cm^3 of 2 mol dm^{-3} sodium hydroxide solution (**Corrosive**) and make up to 100 cm^3 with water.
Label this 'Solution A'.

Solution B: dilute 10 cm^3 of 0.1 mol dm^{-3} potassium hexacyanoferrate(III) solution to 100 cm^3 with water. Label this 'Solution B'.

Teaching tips

Certain envelopes emit a kind of triboluminescence; light is emitted when the glue is pulled apart on opening. This type of self sealing envelope can be purchased cheaply. Opening an envelope in a darkened room gives a surprising flash of light. The envelopes can be resealed and opened again.

Background theory

Students should know that chemiluminescence is light (photons) emitted during the course of a chemical reaction. This is a useful experiment that can be used to practice measurement and how to make up solutions.

Safety

Wear eye protection.

Answers

1. Green necklaces sold at pop concerts, emergency lights for attracting attention when in trouble – *eg* in boats or mountaineering.

An example of chemiluminescence

Introduction

Chemiluminescence is the emission of light during a chemical reaction. In this experiment two solutions are mixed to produce chemiluminescence.

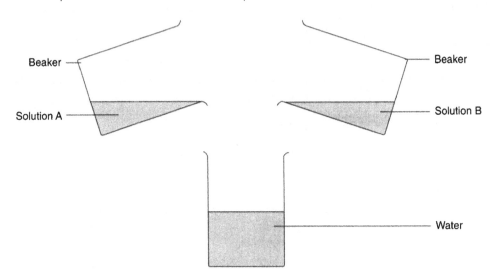

What to do

1. Collect 30 cm^3 of solution (A).
2. Collect 30 cm^3 of solution (B).
3. In a dark room: add 10 cm^3 of each of the two solutions A and B simultaneously to 50 cm^3 of water in a 100 cm^3 beaker.
4. Repeat this but add dyes such as Fluorescein or Methylene blue to the water before mixing A and B.

Safety

Wear eye protection.

Questions

1. Describe one use for chemiluminescence

74. Colorimetric determination of copper ore

Topic

Chemicals from rocks; the extraction of metals.

Timing

60 min.

Description

Students estimate the relative amounts of mineral and waste in a copper ore. The ore is simulated by mixing copper carbonate and sand. Students use a colorimetric method.

Apparatus and equipment (per group)

- ▼ Spatula
- ▼ 100 cm^3 Beaker
- ▼ 250 cm^3 Conical flask
- ▼ Measuring cylinder
- ▼ Filter funnel
- ▼ Filter paper
- ▼ Test-tube rack
- ▼ Six test-tubes
- ▼ Rubber bung
- ▼ Access to balance
- ▼ Access to communal burettes containing the copper(II) sulfate solution and deionised water or two extra 10 cm^3 measuring cylinders and two extra 50 cm^3 beakers.

Chemicals (per group)

- ▼ Simulated copper ore (copper carbonate (**Harmful**) and sand mixture, 30 per cent minimum)
- ▼ Sulfuric acid   2 mol dm^{-3} (**Corrosive**)
- ▼ Copper(II) sulfate solution   1 mol dm^{-3} (**Harmful**)
- ▼ Deionised water.

Teaching tips

The experiment is suitable for students of average ability. Teachers should stress that ores found in nature are not usually this concentrated. For able students, do not give the table in part 8 so that they can calculate the result themselves. They may also use a volumetric flask and take washings in part 4. It is a good idea to set up the standard colours and put white paper at the bottom of the test-tube rack and observe from the top.

Background theory

Understanding the concept of ore and mineral.

RS•C

Safety

Care with sulfuric acid, danger of spray during effervescence.

Answers

1. Copper mineral.
2. Waste; it was removed to enable the colour to be seen properly.
3. Dry and weigh the waste from part 3. Find the weight of mineral by subtraction.

Colorimetric determination of a copper ore

Introduction

An ore is any rock from which a metal may be extracted. Ores contain a mineral of the metal together with waste material. To decide whether an ore is worth mining it is necessary to find out how much of the useful mineral it contains, and how much is waste. This experiment illustrates an example of how this might be done.

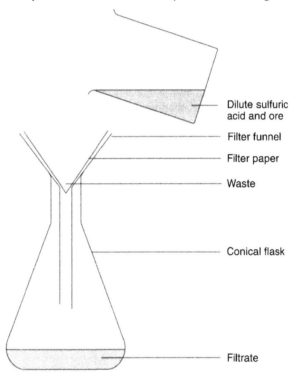

- Dilute sulfuric acid and ore
- Filter funnel
- Filter paper
- Waste
- Conical flask
- Filtrate

What to record

How much copper the ore is estimated to contain.

What to do

1. Weigh 10 g of the ground ore into a beaker.
2. Add 40 cm^3 of 2 mol dm^{-3} sulfuric acid in small amounts. Do not let the mixture go over the top.
3. When the reaction finishes filter the mixture into a conical flask.
4. Add deionised water until the total volume of liquid in the flask is 100 cm^3.
5. Using the laboratory copper(II) sulfate solution, prepare six tubes of diluted copper(II) sulfate, according to the following table. Ensure the solutions are well mixed.

Tube number	1	2	3	4	5
Volume of copper(II) sulfate solution/cm^3	8	6	4	2	0
Volume of deionised water/cm^3	2	4	6	8	10

6. Pour a sample of the solution from your conical flask into another test-tube.
7. Compare the colour of your tube from part 6 with those from part 5. Which one matches the colour best?
8. Estimate the mass of copper mineral in 10 g of the ore using the following table:

Tube of best match	1	2	3	4	5
Mass of compound in 10 g of ore/g	10	7.5	5	2.5	0

Safety

Wear eye protection. Dilute sulfuric acid is corrosive. When gases are made in a reaction, a mist of fine acid spray is often produced which is dangerous to your eyes and causes irritation if inhaled.

Questions

1. Which part of the ore (copper mineral or waste) causes the blue colour of the solutions?
2. Which part of the ore (copper mineral or waste) was removed by filtration in part 3 of the experiment, and why was this done?
3. How could this experiment be adapted to check the result?

75. Glue from milk

Topic

Polymers.

Timing

45 min.

Description

Students make a polymer glue from milk.

Apparatus and equipment (per group)

- 250 cm^3 Beaker
- 250 cm^3 Conical flask
- Stirring rod
- Measuring cylinder
- Paper towels
- Scissors.

Chemicals (per group)

- Skimmed milk
- Ethanoic acid (vinegar)
- Sodium hydrogen carbonate (baking soda NaHCO$_3$)
- Water.

Teaching tips

One way to test the glue is to cut two pieces of paper towel 2.5 cm x 8 cm, stick them end to end using the glue, leave a 2.5 cm overlap. Leave to dry for about 30 min, or until the next lesson. Test by pulling the papers apart lengthways (this is a 'lap shear' test). Depending on the paper used, the paper will tear before the glued section.

Background theory

Students should know the meaning of the term polymer.

Safety

Wear eye protection.

Answers

1. The acid reacts with the protein; casein coagulates. The molecules attract one another forming larger particles, the curds.
2. The sodium hydrogencarbonate neutralises the excess acid.
3. $CH_3COOH(aq) + NaHCO_3(s) \rightarrow CH_3COONa(aq) + H_2O(l) + CO_2(g)$

Glue from milk

Introduction

Glue can be made from the protein in milk called casein. In this experiment, polymer glue is prepared from milk. The casein is separated from milk by processes called coagulation and precipitation.

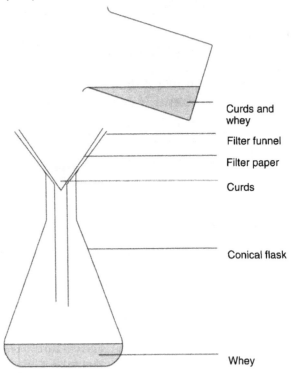

What to do

1. Place 125 cm^3 of skimmed milk into a 250 cm^3 beaker. Add approximately 25 cm^3 of ethanoic acid (or vinegar).
2. Heat gently with constant stirring until small lumps begin to form.
3. Remove from the heat and continue to stir until no more lumps form.
4. Allow the curds to settle, decant some of the liquid (whey) and filter off the remainder using the filter funnel resting on the 250 cm^3 conical flask.
5. Gently remove excess liquid from the curds using the paper towel.
6. Return the solid to the empty beaker. Add 15 cm^3 of water to the solid and stir.
7. Add about half a teaspoon of sodium hydrogen carbonate to neutralise any remaining acid. (Watch for bubbles of gas to appear then add a little more sodium hydrogen carbonate until no more bubbles appear).
8. The substance in the beaker is glue.
9. Find a way to test your glue.

Safety

Wear eye protection.

Questions

1. What is the purpose of the ethanoic acid (vinegar) in this experiment?
2. Why is sodium hydrogencarbonate added?
3. Write an equation for this reaction between ethanoic acid and sodium hydrogencarbonate.

RS•C

76. Rubber band

Topic

Polymers, exothermic reactions, endothermic reactions, thermodynamics, energy.

Timing

30 min.

Description

Students stretch a rubber band and test the effect of heat on a stretched band.

Apparatus and equipment (per group)

- ▼ Rubber band (at least 0.5 cm wide)
- ▼ Hair dryer
- ▼ Weight (>1kg)
- ▼ Ruler.

Teaching tips

The depth of treatment depends on the ability of the students. Students should recognise the difference between exothermic and endothermic reactions. A rubber band width of 1–1.5 cm and a 2 kg mass works well. A ruler standing beside the apparatus is effective as students can see the contraction as it occurs.

Another alternative is to use a clampstand and adjust the height of the weight until it just touches the bench. It is then easy to observe contraction of the rubber band.

Background theory

By placing the rubber band against their lips, students may detect the slight warming that occurs when the rubber band is stretched (exothermic process) and the slight cooling effect that occurs when the rubber band contracts (endothermic process).

The equation $\Delta G = \Delta H - T\Delta S$ (where ΔG means change in Gibb's free energy, ΔH is enthalpy change, ΔS is entropy change and T is the absolute temperature) can be rearranged to give

$T\Delta S = \Delta H - \Delta G$.

The stretching process (exothermic) means that ΔH is negative, and since stretching is non-spontaneous (that is, ΔG is positive and $-\Delta G$ is negative), $T\Delta S$ must be negative. Since T, the absolute temperature, is always positive, we conclude that ΔS due to stretching must be negative. This tells us that rubber under its natural state is more disordered than when it is under tension. When the tension is removed the stretched rubber band spontaneously snaps back to its original shape; that is, ΔG is negative and $-\Delta G$ is positive. The cooling effect means that it is an endothermic process ($\Delta H > 0$), so that $T\Delta S$ is positive. Thus, the entropy of the rubber band increases when it goes from the stretched state to the natural state.

Safety

Ensure rubber bands are sterile and clean. Everyone should have his or her own band. Ensure that if rubber bands break, weights do not drop on toes! Hairdryers should not be brought from home.

Answers

1. Contraction.

2. They should observe that the rubber band contracts when heated, which may well be the opposite of what they have predicted. The most simplistic answer may be that since the endothermic process is favoured when heating occurs, this is a contraction in the case of the rubber polymer since this is the endothermic process.

3.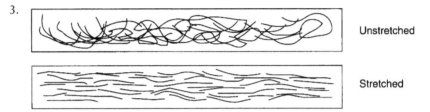

Unstretched

Stretched

Rubber band

Introduction

This experiment involves an investigation into the effect of heat on a stretched rubber band.

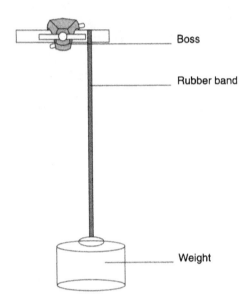

What to record

What was done and what was observed.

What to do

1. Take the rubber band. Quickly stretch it and press it against your lips. Note any temperature change compared with the unstretched band.
2. Now carry out the reverse process. First stretch the rubber band and hold it in this position for a few seconds. Then quickly release the tension and press the rubber band against your lips. Compare this temperature change with the first situation.
3. Set up the apparatus as shown in the diagram. Make sure that if the rubber band breaks the weight cannot drop on toes!
4. Predict what happens if this rubber band is heated with a hair dryer. Write down your prediction. Measure the length of the stretched rubber band.
5. Now heat the rubber band using the hair dryer and observe the result. Does this observation match your prediction? Measure the new length.

Questions

1. Based on your initial testing (by placing the rubber band against your lips) decide which process is exothermic (heat given out): stretching or contracting of the rubber band?
2. The chemist Le Chatelier made the statement '... an increase in temperature tends to favour the endothermic process'. Explain in your own words how this statement and how your answer to question 1 can account for your observations when heating the rubber band.
3. Draw a number of lines to represent chains of rubber molecules showing how they might be arranged in the unstretched and stretched forms. (Hint: the lines of polymer should show less order in the unstretched form than in the stretched form.)

77. Polymer slime

Topic

Polymers, acid base reactions, equilibrium.

Timing

60 min.

Description

Students make a polymer slime and test its properties.

Apparatus and equipment (per group)

- ▼ Plastic teaspoons
- ▼ Plastic or disposable cups (100 cm^3 beakers could be used)
- ▼ Food colouring or fluoroscein dye
- ▼ 100 cm^3 Measuring cylinder
- ▼ Watch glasses or petri dishes
- ▼ Two dropping pipettes. Use the type of teat pipette (usually fitted to Universal Indicator bottles) that do not allow squirting. – eg Griffin.

Chemicals (per group)

- ▼ Polyvinyl alcohol solution (4 per cent)
- ▼ Borax (sodium tetraborate) solution 4 per cent (one quarter the volume of polyvinyl alcohol solution).
- ▼ Sodium hydroxide solution   0.4 mol dm^{-3} (**Irritant**)
- ▼ Hydrochloric acid   0.4 mol dm^{-3} (**Irritant**).

Polyvinyl alcohol solution

To prepare the polyvinyl alcohol solution use hydrolysed polyvinyl alcohol powder (available from BDH). To prepare a 4 per cent solution, weigh 40 g of hydrolysed polyvinyl alcohol. Gradually heat 1 dm^3 of tap water to 50 °C, then gradually sprinkle the solid polymer across the surface, while stirring constantly (best with a hot plate magnetic stirrer). Continue heating to 90 °C gradually, and stirring. Do not overheat or boil the solution. The solution should appear colourless and clear at this point, with practically all solids dissolved. Remove from heat; cover with aluminium foil and allow to cool overnight. The solution can be kept for quite a period of time stored in plastic bottles.

Borax solution

Prepare a 4 per cent w/v solution in water. Calculate what total volume of solution is required for the class to weigh out the appropriate amount of borax. The student should add approximately one quarter of the volume of the polyvinyl alcohol solution.

RS•C

Background theory

Polyvinyl alcohol is an addition polymer formed from vinyl alcohol. The polymer can dissolve in water as a result of the OH (hydroxyl) groups attached to the main polymer chain, which form hydrogen bonds with water molecules.

It is helpful if students understand:

how polymers are formed from monomers, the concept of equilibrium and that an equilibrium exists

$$B(OH)_3 + 2H_2O \rightleftharpoons B(OH)_4^- + H_3O^+$$

The $B(OH)_4^-$ ion is believed to crosslink the polymer chains as shown

Safety

Wear eye potection. The polyvinyl alcohol and borax solutions are non-toxic, but wash your hands when finished. Students should not take slime home.

Answers

1. Slime stretches.
2. Slime breaks.
3. Slime bounces.
4. Slime does not break, but instead thickens under stress.
5. Slime dissolves water-based ink, so writing is taken up on the slime.
6. The acid destroys the properties of the gel because bonds between the chain are broken.
7. The base restores the properties of the gel because it neutralises the acid allowing the crosslinking to occur again. Discussion of acid base equilibria can be involved here, since tests 6 & 7 can be repeated and the same results obtained.

Polymer slime

Introduction

A solution of polyvinyl alcohol can be made into a gel (slime) by adding a borax solution, which creates crosslinks between chains. In this activity, some interesting properties of the slime are investigated.

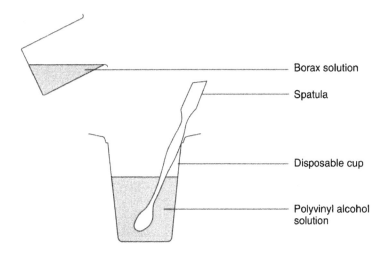

What to record

Results of the tests.

What to do

1. Collect 40 cm^3 of polyvinyl alcohol solution in a disposable cup containing a spatula.
2. If desired add one drop of food colour or fluoroscein dye to the solution. Stir well.
3. Measure 10 cm^3 of borax solution and add this to the polyvinyl alcohol solution. Stir vigorously until gelling is complete.
4. Remove the slime from the cup and pat and knead it thoroughly to completely mix the contents. Roll the slime around in your hand, gently squeezing the material to remove air bubbles at the same time.

Safety

Wear eye protection.

Questions

Test the properties of your slime

1. Pull slowly – what happens?
2. Pull sharply and quickly – what happens?
3. Roll the slime into a ball and drop it on the bench – what happens?
4. Place a small bit on the bench and hit it with your hand – what happens?
5. Write your name on a piece of paper with a water based felt tip pen. Place the slime on top, press firmly, then lift up slime. What happened to the writing? To the slime? Try the same thing using a spirit-based pen. Does this show the same effect?

6. Place a small piece of your slime on a watch glass or petri dish. Add dilute hydrochloric acid dropwise, stirring well after each drop. When a change is noticed record the number of drops added and your observations.
7. Now add dilute sodium hydroxide solution dropwise to the same sample used in 6 stirring after each drop. When a change is noticed record your observations. Can the whole process be repeated with tests 6 and 7? Try it!

78. The properties of ethanoic acid

Topic

Organic chemistry.

Timing

30 min.

Description

Some courses require knowledge of the simple properties of organic acids. This experiment illustrates the properties of ethanoic acid as a typical weak acid.

Apparatus and equipment (per group)

- ▼ Test-tubes
- ▼ Stirring rod.

Chemicals (per group)

- ▼ Access to dilute ethanoic acid $\quad$ 0.05 mol dm^{-3}
- ▼ Sodium carbonate solution or sodium hydrogencarbonate solution $\quad$ 0.4 mol dm^{-3}
- ▼ Sodium hydroxide solution $\quad$ 0.4 mol dm^{-3}
- ▼ Small piece of magnesium ribbon (about 2 cm long cleaned to remove any oxide coating)
- ▼ Full range indicator solution.

Teaching tips

One or two drops of sodium hydroxide turns the indicator alkali. More drops of sodium carbonate are needed. Magnesium reacts with the ethanoic acid forming hydrogen.

Background theory

Students need to appreciate the simple idea of functional groups – *ie* that other organic acids behave in a similar way.

Safety

Wear eye protection.

Answers

1. (a) $2HCl + Na_2CO_3 \rightarrow 2NaCl + H_2O + CO_2$
   (b) $HCl + NaOH \rightarrow NaCl + H_2O$
   (c) $Mg + 2HCl \rightarrow MgCl_2 + H_2$
2. (a) $2CH_3CO_2H + Na_2CO_3 \rightarrow 2CH_3CO_2Na + CO_2 + H_2O$
   (b) $CH_3CO_2H + NaOH \rightarrow CH_3CO_2Na + H_2O$
   (c) $2CH_3CO_2H + Mg \rightarrow Mg(CH_3CO_2)_2 + H_2$

RS•C

The properties of ethanoic acid

Introduction

Acids are an important group of chemicals. Organic acids are characterised by the presence of a -COOH group attached to a carbon atom. In this experiment, some typical properties of a weak organic acid are observed.

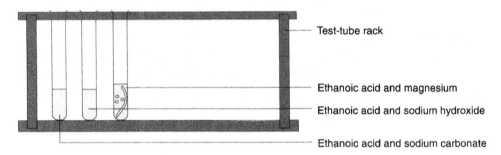

- Test-tube rack
- Ethanoic acid and magnesium
- Ethanoic acid and sodium hydroxide
- Ethanoic acid and sodium carbonate

What to record

What was observed.

What to do

1. Use a small quantity (1–2 cm^3 at a time) of dilute ethanoic acid.
2. Observe the effect on full range indicator paper of adding drops of sodium carbonate solution to 2 cm^3 of dilute ethanoic acid.
3. Repeat the experiment with dilute sodium hydroxide solution.
4. Add a small piece of magnesium ribbon to the ethanoic acid. What is observed? Try to confirm the identity of the gas given off.

Safety

Wear eye protection.

Questions

1. Write equations for the reactions of hydrochloric acid (HCl) with: -
    (a) sodium carbonate solution (Na$_2$CO$_3$)
    (b) sodium hydroxide solution (NaOH)
    (c) magnesium ribbon (Mg).
2. The formula of ethanoic acid is written as CH$_3$COOH. Write similar equations for the reactions of ethanoic acid with:
    (a) sodium carbonate solution
    (b) sodium hydroxide solution
    (c) magnesium.

79. Properties of alcohols

Topic

Organic chemistry.

Timing

30 min.

Description

In this experiment students note the miscibility of ethanol and water, and test the pH. Students burn some alcohol and carry out an oxidation of ethanol to produce ethanaldehyde.

Apparatus and equipment (per group)

- ▼ Test-tubes
- ▼ Tin lid.

Chemicals (per group)

- ▼ Ethanol (**Highly flammable**)
- ▼ Potassium dichromate(VI) solution   0.01 mol dm^{-3} (**Toxic**)
- ▼ Dilute sulfuric acid   1 mol dm^{-3} (**Irritant**)
- ▼ Full range indicator solution.

Teaching tips

Students may not have come across the concept of refluxing to prevent losing reagents. For this reason it is not introduced in this activity. Teachers may wish to demonstrate the method. The alcohol is oxidised to an aldehyde, which boils off as it has a lower boiling point than the alcohol. If the reaction is heated under reflux further oxidation to the acid occurs.

Ethanol mixes with water and it has pH 7. It burns with a light blue flame. The oxidative solution turns from an orange to a blue-green as aldehyde forms.

Background theory

Students need to appreciate a simple idea of functional groups – *ie* that other alcohols behave in a similar way.

Safety

Wear eye protection. Potassium dichromate can cause ulcers on contact with the skin. Wear disposable gloves. Keep stocks of ethanol away from naked flames.

Show students how to smell safely.

Answers

1. Ethanol is made by fermentation.
2. Reduction.
3. Ethanol + oxygen → carbon dioxide + water

Properties of alcohols

Introduction

Alcohols are an important group of organic chemicals. The alcohol people drink is called ethanol and is produced by fermentation. Alcohols are characterised by an -OH group attached to a carbon atom.

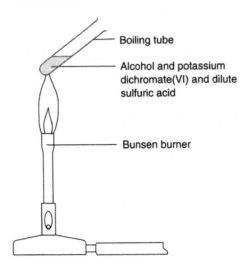

What to record

What was observed.

What to do

1. Take a small quantity of ethanol and add the same volume of water. What is the pH of the mixture? Test the mixture with full range indicator solution. Does the water mix with the ethanol?
2. Put a small quantity of ethanol on a tin lid and ignite it with a splint. Does it burn, and if so, describe the flame.
3. Put 5 cm^3 of dilute sulfuric acid in a boiling tube. Add five drops of potassium dichromate(VI) solution. Now add two drops of ethanol and heat the mixture until it just boils. Is there any sign of a reaction? Is there any change of smell that could come from a new compound?

Safety

Ethanol is highly flammable. Potassium dichromate (VI) is toxic. Wear disposable gloves. Avoid skin contact. Wash hands after use.

Questions

1. What is the name of the process used to produce ethanol on a commercial scale?
2. The reaction of an alcohol to produce an acid is called an oxidation reaction. What is the opposite reaction called that produces an alcohol from an acid?
3. Write a chemical equation for the combustion of ethanol.

80. Testing salts for anions and cations

Topic

Qualitative analysis.

Timing

1–2 hours.

Description

Students attempt to identify the anions and cations present in a salt by a combination of tests.

Apparatus and equipment (per group)

▼ Test-tubes.

Chemicals (per group)

Access to:

▼ Full range indicator paper
▼ Ammonia solution — 2 mol dm^{-3}
▼ Sodium hydroxide solution — 0.4 mol dm^{-3} (**Irritant**)
▼ Hydrochloric acid solution — 0.4 mol dm^{-3}
▼ Barium chloride solution — 0.1 mol dm^{-3} (**Harmful**)
▼ Limewater solution — 0.02 mol dm^{-3}
▼ Nitric acid — 0.4 mol dm^{-3} (**Irritant**)
▼ Silver nitrate solution — 0.1 mol dm^{-3}
▼ Unknown substances labelled A, B, C …each might contain one of the following anions and one of the following cations:

Anions - OH^-, SO_4^{2-}, CO_3^{2-}, Cl^-, Br^-, I^-, NO_3^-
Cations - H^+, Ca^{2+}, Cu^{2+}, Fe^{3+}, Fe^{2+}, NH_4^+

A sensible selection might be:

copper chloride (**Toxic**), potassium carbonate, potassium iodide, copper(II) sulfate (**Harmful**), iron(III) chloride (**Irritant**), iron(II) sulfate, lead nitrate (**Toxic**).

Extension – for flame tests

Nichrome wire loops attached to wooden handles (cleaned before lesson in concentrated hydrochloric acid).

Safety

Wear eye protection. Ammonia solution causes burns and gives off ammonia vapour which irritates the eyes, lungs and respiratory system.

Sodium hydroxide can cause burns and is dangerous to the eyes.

Hydrochloric acid can cause burns.

Barium chloride is harmful by inhalation and if swallowed.

Nitric acid causes burns.

Silver nitrate solution causes burns.

Teaching tips

Test-tubes should be washed initially. Thorough washing to prevent contamination is important.

Flame tests
It is probably inadvisable to use concentrated hydrochloric acid to produce volatile chlorides at this level. This procedure should be effective as long as sodium, which produces a persistent yellow colour, is not given as an unknown.

For another method of flame test demonstration, see *Classic Chemistry Demonstrations*, p. 80. London: RSC, 1995.

This experiment is probably suitable for able 15/16 year old students in this format. Teachers may wish to adapt this for less able students and or spread the work over 2 or 3 lessons.

Background theory

A knowledge of precipitation reactions is helpful as is pre-knowledge of the chemistry of the tests. Otherwise, the students should test known substances to ensure they know what is a positive result.

Testing salts for anions and cations

Introduction

Chemists often have to identify the composition of unknown substances. This experiment involves identifying the cations and anions in various salt solutions.

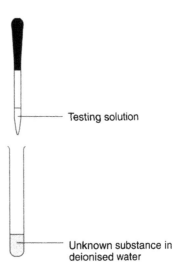

What to record

Sample	Test performed	Result of test

What to do

1. Dissolve the unknown substance in deionised water. 5–10 cm^3 of solution may be needed.
2. Using the analysis table, test small aliquots (portions).
3. Repeat for the other unknown substances.

Safety

Wear eye protection. Some of the unknowns may be toxic or corrosive.

Questions

1. Write word and ionic equations for those reactions that give a positive result.

Testing salts for anions and cations.

For anions: carry out the three tests A, B and C below:

Test	Anion	Test and observation
A Silver nitrate followed by ammonia solution	Chloride (Cl^-)	Add a few drops of dilute nitric acid (**Irritant**) followed by a few drops of silver nitrate solution. A white precipitate of silver chloride is formed. The precipitate is soluble in ammonia solution.
	Bromide (Br^-)	Add a few drops of dilute nitric acid (**Irritant**) followed by a few drops of silver nitrate solution. A pale yellow precipitate of silver bromide is formed. The precipitate is slightly soluble in ammonia solution.
	Iodide (I^-)	Add a few drops of dilute nitric acid followed by a few drops of silver nitrate solution. A yellow precipitate of silver iodide is formed. It is insoluble in ammonia solution.
B Barium chloride	Sulfate (SO_4^{2-})	Add a few drops of barium chloride solution (**Toxic**) followed by a few drops of dilute hydrochloric acid. A white precipitate of barium sulfate is formed.
C Hydrochloric acid	Carbonate (CO_3^{2-})	Add dilute hydrochloric acid to the solution (or add it to the solid). Bubbles of carbon dioxide are given off.

For cations: carry out the two tests D and E below:

Cation	D Add sodium hydroxide solution (Irritant)	E Add ammonia solution
Ammonium (NH_4^+(aq))	Warm carefully. Do not allow to spit. Ammonia (alkali gas) is given off	—
Copper (Cu^{2+}(aq))	Blue (jelly-like) precipitate of $Cu(OH)_2$(s)	Blue jelly like precipitate dissolves in excess ammonia to form a deep blue solution.
Iron(II) (Fe^{2+}(aq))	Green gelatinous precipitate of $Fe(OH)_2$(s)	Green gelatinous precipitate
Iron(III), (Fe^{3+}(aq))	Rust-brown gelatinous precipitate of $Fe(OH)_3$(s)	Rust brown gelatinous precipitate
Lead(II), (Pb^{2+}(aq))	White precipitate $Pb(OH)_2$(s) dissolves in excess NaOH(aq)	White precipitate, $Pb(OH)_2$
Zinc (Zn^{2+}(aq))	White precipitate, $Zn(OH)_2$(s)	White precipitate, $Zn(OH)_2$(s) dissolves in excess NH_3(aq)
Aluminium (Al^{3+}(aq))	Colourless precipitate, $Al(OH)_3$(s)	Colourless precipitate, $Al(OH)_3$(s)

Flame tests.

1. Slightly open the air hole of the Bunsen burner.
2. Heat a piece of nichrome wire in a Bunsen flame until the flame is no longer coloured.
3. Dip the loop at the end of the wire into some water.
4. Dip the loop into an unknown salt.
5. Hold the wire in the edge of the flame.
6. Record the colour and identify the cation using the table below.

Metal	Colour of flame
Barium	Apple-green
Calcium	Brick-red
Copper	Green with blue streaks
Lithium	Crimson
Potassium	Lilac
Sodium	Yellow

RS•C

81. Quantitative electrolysis

Topic

The mole and electrolysis.

Timing

60 min.

Description

Students perform a quantitative electrolysis using copper electrodes and copper(II) sulfate solution. The number of moles of metal removed is calculated, and linked to the amount of electricity passed through the circuit.

Apparatus and equipment (per group)

- ▼ 6 V DC supply
- ▼ Ammeter 0–1 A
- ▼ Five connecting leads
- ▼ 100 cm^3 Beaker
- ▼ Access to a balance reading to 0.01 g.

Chemicals (per group)

- ▼ Two copper foil electrodes (approximately 5 cm x 3 cm)
- ▼ Copper(II) sulfate solution 0.1 mol dm^{-3}

Teaching tips

Use reduction in mass of the anode to explain the principles. Copper can easily fall off the cathode and it can be wiped off when drying (especially when deposited at higher currents or when the cathode is dirty).

The electrodes should be cleaned before use.

For an indication of the theoretical loss of mass: 0.4 A for 30 min removes
0.4 x 30 x 60 x 0.5 x 63.5 = 0.237 g of copper.

Current should be kept constant by direct adjustment of the power supply or by the inclusion of a rheostat to the circuit. It has been suggested that adjusting the position of the electrodes is also effective.

Background theory

Students require knowledge of charges on ions, the mole, and the relationship between quantity of charge, current and time. They can also be introduced to the idea of a mole of electrons.

Safety

Wear eye protection.

Answers

1,2,3 will depend on experimental results.

4. Main errors are fluctuations in the current and inaccurate measurements.

Quantitative electrolysis

Introduction

When electrolysis is done on a commercial scale it is important to know how much current is required and for how long. This experiment relates the amount of metal removed from an electrode to the electric current and the time the current flows.

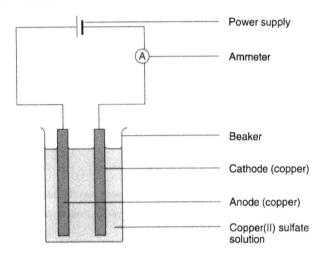

What to record

▼ The masses of the electrodes before electrolysis (identify the electrodes by writing on them).
▼ The masses of the electrodes after electrolysis.
▼ The current flowing.
▼ The time the current flows.

What to do

1. Clean the electrodes with emery paper (avoid inhaling any dust).
2. Weigh the anode.
3. Immerse the electrodes to a depth of 3–4 cm in the solution of copper(II) sulfate.
4. Allow about 0.4 A to pass for about 30 min.
5. Remove the anode, wash carefully in water and dry gently with a paper towel.
6. Reweigh the anode.

Safety

Wear eye protection.

Questions

1. Calculate the number of moles of copper that have been removed from the anode.(Mass lost in g / 63.5)
2. Calculate the charge that has flowed through the circuit using the relationship charge (in Coulombs) = current (in amps) x time (in seconds).
3. Using the answers to questions 1 and 2, calculate the number of Coulombs required to remove one mole of copper.
4. 193,000 (2 x 96,500) Coulombs is required to remove one mole of copper. The difference between this and the answer to question 3 is due to errors in the experiment. What are the main sources of error in this experiment?

82. The electrolysis of solutions

Topic

Electrolysis of aqueous solutions leading to the ability to predict results, reactivity series.

Timing

60 min.

Description

Students electrolyse some aqueous solutions and identify the products.

Apparatus and equipment (per group)

- 100 cm^3 Beaker
- Electrodes (S-shaped are preferable)
- Small test-tubes for gas collection
- Two leads
- Two crocodile clips
- DC power supply (6 V is reasonable)
- Eye protection.

Chemicals (per group)

- Universal Indicator paper
- Access to solutions of :
- Sodium chloride   0.5 mol dm^{-3}
- Copper chloride   0.5 mol dm^{-3} (**Harmful**)
- Potassium iodide   0.5 mol dm^{-3}
- Sodium bromide   0.5 mol dm^{-3}
- Potassium sulfate   0.5 mol dm^{-3}
- Copper(II) sulfate   0.5 mol dm^{-3}
- Lead nitrate   0.5 mol dm^{-3} (**Toxic**)
- Silver nitrate   0.1 mol dm^{-3}.

Teaching tips

Teachers may wish to show students how to fill and invert the small test-tubes over the electrodes. With care, these can be inverted without spilling the liquid. Using small tubes filled with water rather than the test solution is safer. These are then inverted into the solution to be tested. The tubes may be clamped or supported by rubber bands wound round the test-tubes and round the electrode. The class results can be pooled as there will not be time to test all the solutions individually.

Students may need reminding of, or introducing to, the common tests for hydrogen, oxygen, chlorine, bromine and iodine.

Solution	Product at anode	Product at cathode
Sodium chloride	Chlorine	Hydrogen
Copper chloride	Chlorine	Copper
Potassium iodide	Iodine	Hydrogen
Sodium bromide	Bromine	Hydrogen
Potassium sulfate	Oxygen	Hydrogen
Copper(II) sulfate	Oxygen	Copper
Lead nitrate	Oxygen	Lead
Silver nitrate	Oxygen	Silver

Background theory

Students should know what happens when molten ionic solids are electrolysed. If students do not know the tests to identify the products, this is an appropriate time to introduce them.

Safety

Wear eye protection.

Chlorine is toxic and harmful to the lungs, eyes and respiratory tract.

Bromine vapour is an irritant and very toxic if inhaled.

Iodine is harmful by skin contact.

Hydrogen is extremely flammable.

Oxygen supports combustion.

Ensure good ventilation. Do not allow chlorine or bromine vapour to be produced for very long.

Answers

1. Metals or hydrogen.
2. Non-metals.
3. From the water.
4. (a) The metal is produced if it is lower than hydrogen in the reactivity series, otherwise hydrogen is produced.
   (b) Halides give halogens, sulfates and nitrates give oxygen.

The electrolysis of solutions

Introduction

When electricity passes through molten compounds, like sodium chloride, the ions move towards the electrode of opposite charge. Sodium chloride gives sodium metal and chlorine gas. This experiment illustrates what happens when the system is made more complicated because water is present. Electricity is passed through various solutions and the products are identified.

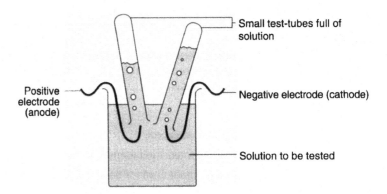

What to record

Solution	Product at the anode	Product at the cathode

What to do

1. Set up the apparatus as shown.
2. Switch on and observe what happens.
3. Try to identify the gases produced (if any).

Safety

Wear eye protection

The gases produced may be flammable, oxidising, and toxic. Take care not to inhale them. Do not let the current flow for very long. Some of the solutions are toxic.

Questions

1. What type of element is formed at the negative electrode?
2. What type of element is formed at the positive electrode?
3. Your table of results should show some products, which could not come from the compound itself that was electrolysed. Where could these other products have come from?
4. Write a general rule for the products formed at
   (a) the cathode
   (b) the anode.

83. An oxidation and reduction reaction

Topic

Oxidation-reduction.

Timing

15 min.

Description

This experiment illustrates oxidation and reduction. Methylene blue is reduced by an alkaline dextrose solution to produce a colourless solution. When this solution is shaken, it is oxidised by the oxygen in the flask to produce the blue dye.

Apparatus and equipment (per group)

- ▼ 250 cm^3 Conical flask
- ▼ Stopper to fit.

Chemicals (per group)

- ▼ Potassium hydroxide solution   27 g dm^{-3} (**Irritant**)
- ▼ Dextrose (glucose)
- ▼ A few drops of methylene blue indicator.

Teaching tips

Allow students to guess that the oxygen in the flask is causing the reaction to occur. For another method and references see *Classic Chemistry Demonstrations*, p. 48. London: RSC, 1995.

Background theory

The reaction occurs in a number of steps. Students can be taught:

Gas + liquid → blue colour

And clearing:

Blue colour + alkaline dextrose → colourless

Students can guess the gas is oxygen.

Safety

Wear eye protection. Check the stopper is a good fit and does not leak when using water.

Answers

1. To introduce more oxygen into the flask.

An oxidation and reduction reaction

Introduction

A conical flask contains a colourless solution. When shaken, a blue colour forms. After a few seconds, the blue colour fades and the solution again becomes colourless. The process can be repeated. It is an oxidation followed by a reduction process

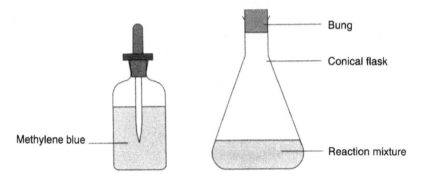

What to do

1. Put some water in the conical flask. Put in the stopper. Shake vigorously to check for leaks. If there are none, pour the water away and proceed.
2. Put 100 cm^3 of potassium hydroxide solution into a conical flask.
3. Add 3.3 g dextrose.
4. Add 3–4 drops of methylene blue indicator.
5. Put a stopper on the flask.
6. Shake vigorously.
7. When the solution clears, repeat the process.
8. It is necessary periodically to remove the stopper.

Safety

Wear eye protection.

Questions

1. Why is it necessary to periodically remove the stopper?

Classic chemistry experiments

RS•C

84. Heats of reaction (exothermic or endothermic reactions)

Topic

Energy transfer in chemical reactions, exothermic and endothermic reactions.

Timing

60 min.

Description

Students investigate the heat absorbed or evolved on mixing chemicals together. They can calculate the heat absorbed per mole to compare relative amounts.

Apparatus and equipment (per group)

- ▼ Eye protection
- ▼ Four test-tubes or four expanded polystyrene cups with lids to act as calorimeters
- ▼ Spatula
- ▼ Teat pipette or small measuring cylinder
- ▼ Thermometer
- ▼ Access to a balance.

Chemicals (per group)

- ▼ Anhydrous copper(II) sulfate (**Harmful**)
- ▼ Citric acid crystals (2-hydroxy-1,2,3-propane tricarboxylic acid)
- ▼ Sodium hydrogencarbonate
- ▼ Copper(II) sulfate solution 0.5 mol dm^{-3}
- ▼ Zinc powder (**Flammable**).

Teaching tips

Students may be surprised that energy can either be evolved or absorbed in reactions. To make a chemical bond; another bond must first be broken. It is the sum of the energy changes in making and breaking bonds that results in the overall energy change. If temperature sensors and data logging equipment are available, they may be appropriate in this context.

A temperature sensor attached to a computer can be used in place of a thermometer. It can plot the temperature change on a graph and make a helpful demonstration of what happens when chemicals react. This data logging set up might be the basis for a project where students have to find the mix of chemicals that yield the optimal heat loss or gain.

Background theory

To carry out mole calculations students need to be able to translate reacting quantities in grams into moles, and to be able to relate energy changes to the mass of solution (water) used x specific heat capacity x temperature change. If students are to attempt this, they need to weigh the reactants rather than using spatula measures. Energy level

RS•C

diagrams help in understanding that the evolution of energy leads to a loss in energy of the system. Exothermic change is given a negative sign.

Safety

Wear eye protection.

Answers

1. Experiments 1 and 3 are exothermic, experiment 2 is endothermic.
2. Copper(II) sulfate + zinc → zinc sulfate + copper
   $CuSO_4 + Zn \rightarrow ZnSO_4 + Cu$
3. Citric acid and sodium hydrogen carbonate.
4. Anhydrous copper(II) sulfate and water, or copper(II) sulfate solution and powdered zinc.

Heats of reaction (exothermic or endothermic reactions)

Introduction

Instant hot and cold packs are available for use in first aid. This experiment illustrates the types of chemical reaction that occur in these packs.

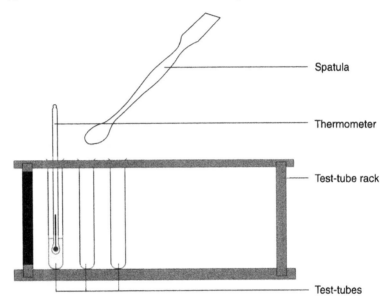

What to record

What was done and any changes in temperature from the starting temperature of your reaction. A table may be useful.

Initial solution	Temperature of solution/°C	Solid added	Final temperature of mixture/°C	Temperature change	Type of reaction

What to do

Experiment 1.
1. Put 2 cm^3 of water in a test-tube.
2. Record the temperature of the water.
3. Add a spatula measure of anhydrous (white) copper (II) sulfate.
4. Carefully stir, using the thermometer, and record the temperature again.

Experiment 2.
1. In a dry test-tube mix one spatula measure of citric acid and one spatula measure of sodium hydrogencarbonate.
2. Put 2 cm^3 of water in another test tube.

RS•C

3. Record the temperature of the water.
4. Add the mixture to the water.
5. Watch what happens and take the temperature of the solution.

Experiment 3.
1. Put about 5 cm^3 of copper(II) sulfate solution in a test-tube.
2. Record the temperature.
3. Add a spatula measure of powdered zinc.
4. Record the new temperature.

Safety

Wear eye protection.

Anhydrous copper(II) sulfate is harmful.

Zinc powder is flammable.

Questions

1. Which reactions are exothermic and which are endothermic?
2. Write word and symbol equations for Experiment 3.
3. Which two substances could be put in a cold pack?
4. Golfers need a hand warmer to keep their hands warm on a cold day. Which chemicals could be put in these warmers?

85. Comparing the heat energy produced by combustion of various alcohols

Topic

Exothermic reactions, bond formation, bond breaking, heats of reaction/combustion.

Timing

60 min.

Description

Students use the combustion of known amounts of various alcohols to heat a known amount of water.

Apparatus and equipment (per group)

- ▼ Retort stand
- ▼ Boss
- ▼ Clamp
- ▼ 250 cm^3 Conical flask or metal calorimeter
- ▼ Thermometer
- ▼ Spirit burners.

Chemicals (per group)

Methanol (**Highly flammable**)
Ethanol (**Highly flammable**)
Propanol (propan-1-ol) (**Highly flammable**)
Butanol (butan-1-ol) (**Highly flammable**).

Teaching tips

Stress the importance of fair testing, for example the height of the calorimeter above the wick. More able students can work out the number of moles used and find the energy produced per mole. A temperature sensor attached to a computer can be used in place of a thermometer. It can plot the temperature change on a graph and allow better quantification of the heat produced.

Background theory

Energy changes, exothermic and endothermic reactions, reactions involving bond breaking and bond formation. For more advanced students bond enthalpies.
Energy change = mass of solution x specific heat capacity x temperature change.

Safety

Wear eye protection. Spirit burners are the safest way of burning the alcohol but students should not be allowed to fill them. Use spirit burners with a wide base. Spirit burners should be filled by a technician in the preparation room and labelled. No stock bottles of alcohol should be allowed in the laboratory.

Answers

1. Butanol.

Comparing the heat energy produced by combustion of various alcohols

Introduction

The combustion of alcohol produces energy. This experiment compares the amount of heat produced by the combustion of various alcohols.

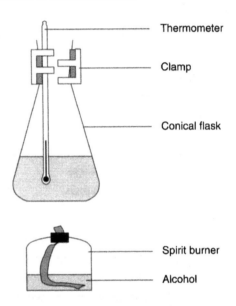

What to record

Alcohol	Initial temp/°C	Final temp/°C	Temp change/°C	Initial mass/g	Final mass/g	Mass used/g
Methanol						
Ethanol						
Propanol						
Butanol						

What to do

1. Fill the conical flask with 100 cm^3 of water. Clamp the flask at a suitable height so that the spirit burner can be easily placed below.
2. Weigh the spirit burner (and lid) containing the alcohol and record the mass and name of the alcohol.
3. Record the initial temperature of the water using the thermometer.
4. Place the spirit burner under the conical flask and light the wick.

5. Allow the alcohol to heat the water so the temperature rises by about 40 °C.
6. Replace the cap to extinguish the flame.
7. Reweigh the spirit burner and cap and work out the mass of alcohol used.

Repeat for different alcohols. Use 100 cm^3 of new cold water each time.

Safety

Wear eye protection. Do not open the spirit burner.

Question

1. Which fuel provides the most energy per gram?

RS•C

86. Fermentation

Topic

Enzymes, rates, synthesis, distillation.

Timing

Two lessons.

Description

Ethanol is produced from the fermentation of glucose by yeast. Students leave a sugar and yeast solution to ferment between lessons. The students test for carbon dioxide and note the characteristic smell of the alcohol produced.

Apparatus and equipment (per group)

- ▼ 100 cm^3 Conical flask
- ▼ Boiling tube
- ▼ Cotton wool
- ▼ Sticky labels.

Chemicals (per group)

- ▼ Glucose
- ▼ Any fast acting yeast
- ▼ Limewater 0.02 mol dm^{-3}
- ▼ Warm water 30–40 °C.

Teaching tips

Pool class results in second lesson and demonstrate distillation. Different groups could be asked to do the experiment at different temperatures and the rate of production of bubbles compared. Hence introduce the idea that there is an optimum temperature for enzyme activity.

Fermentation by yeast produces changes which can be easily monitored using a data logging system. Of these changes pressure sensors can measure gas evolved, while pH sensors respond to increases in dissolved carbon dioxide.

Background theory

Students need to know the limewater test for carbon dioxide.

Safety

Wear eye protection. Do not let students taste the alcohol. Laboratory apparatus and chemicals may well be contaminated.

Answers

1. Carbon dioxide.
2. Mass lost, rate of bubble production, gas syringe.
3. Glucose $\xrightarrow{zymase}$ ethanol + carbon dioxide

Fermentation

Introduction

Beer and wine are produced by the fermentation of glucose by yeast. In this experiment, a glucose solution is left to ferment. The resulting mixture is then tested for the presence of any fermentation products.

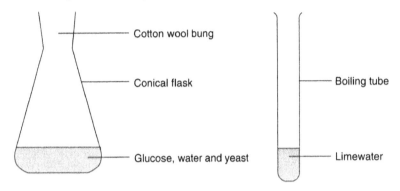

What to record

What was seen to happen to the limewater.

What to do

Lesson 1
1. Put 5 g of glucose in the flask and dissolve it in 50 cm^3 of warm water.
2. Add 1 g of yeast to the glucose solution and plug the top with cotton wool.
3. Wait 20 min while fermentation takes place.
4. Remove the cotton wool and pour the invisible gas from the flask into the boiling tube with limewater. Take care not to pour any of the liquid as well.
5. Gently swirl the limewater round the tube and observe what happens.

Lesson 2
1. Note the smell of the solution.

Safety

Wear eye protection. Do not taste the product.

Questions

1. What gas is present in the flask after fermentation?
2. Suggest a different method for measuring the speed of this reaction.
3. Yeast contains a chemical called zymase, which is an enzyme. Complete the word equation for fermentation.

$$\text{Glucose} \xrightarrow{\text{zymase}} \qquad +$$

RS•C

87. Microbes, milk and enzymes

Topic

Rates, microbes and enzymes.

Timing

One hour, but over two lessons to allow for incubation.

Description

Students test various different types of cow's milk and determine the number of bacteria present by the colour of resazurin.

Apparatus and equipment (per group)

- Test-tube rack
- Five test-tubes
- Sticky labels
- 250 cm^3 Beaker
- Tripod
- Gauze
- Bunsen burner.

Chemicals (per group)

- UHT milk
- Pasteurised full fat milk
- Pasteurised semi-skimmed milk
- Pasteurised skimmed milk
- Powdered milk
- Resazurin solution (this should be freshly prepared).

Teaching tips

Students should be aware of the need for clean apparatus. A similar experiment may be carried out using methylene blue.

Background theory

Students should know that rates of reaction are influenced by a variety of factors and that enzymes produced by microbes are biological catalysts.

Safety

Wear eye protection. Warn students not to taste their milk as it may contain pathogenic microbes. Laboratory apparatus may be contaminated.

Microbes, milk and enzymes

Introduction

Microbes are responsible for the production of some foods, for example cheese and yoghurt. They are also responsible for food decay. The enzymes they contain are catalysts. This experiment shows how these microbes and enzymes effect various types of milk.

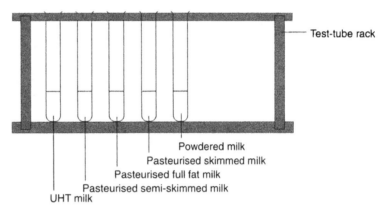

Powdered milk
Pasteurised skimmed milk
Pasteurised full fat milk
Pasteurised semi-skimmed milk
UHT milk

What to do

The experiment needs to be done over two lessons.

First lesson.
1. Place five test-tubes in a test-tube rack. Fill each one to a depth of about 3 cm with five different types of milk.
2. Label each type with a sticky label near the top of the test-tube.
3. Leave in a warm place for between 3-5 days.

Second lesson.
1. Fill a beaker with about 100 cm^3 of tap water and stand the test-tubes in the beaker. Heat over a Bunsen burner to about 60 °C.
2. Turn off the Bunsen burner and carefully lift the beaker off the tripod.
3. Put six drops of rezasurin indicator in each test-tube of milk, shake thoroughly with the normal side to side action.
4. Leave to stand for 15 min and note any colour change.

Key

The rezasurin indicates the number of bacteria present.

Colour	Number of bacteria	Condition of milk
Purple	None	Completely sterilised
Blue	Few	Milk still fresh
Pink	Some	Milk on the turn
Colourless	Many	Milk has gone off

Safety

Do not taste any of the milk.

RS•C

88. Properties of the transition metals and their compounds

Topic

The periodic table, transition metals.

Timing

60 min.

Description

Students extend their knowledge of the Periodic Table by examining the transition metals and their compounds.

Apparatus and equipment (per group)

- ▼ Test-tubes
- ▼ Access to a bar magnet
- ▼ Dropping pipette. Use the type of teat pipette (usually fitted to Universal Indicator bottles) that does not allow squirting – *eg* Griffin.

Chemicals (per group)

- ▼ Samples of some transition metals (copper, iron, nickel, zinc)

Access to solutions of:

▼ Copper(II) sulfate	0.01 mol dm^{-3}
▼ Iron(III) chloride	0.1 mol dm^{-3} (**Irritant**)
▼ Nickel(II) chloride	0.1 mol dm^{-3}

or other compounds with similar oxidation states

- ▼ Ammonia solution   2 mol dm^{-3}

As many solid samples of transition metal compounds as possible in closed containers for observation of colours.

Teaching tips

This experiment is a good test of observational skills, and students' attention could be drawn to this. In the reaction with water very little happens and when forming the complexes some colour changes could be missed.

If students have not used an inverted filter funnel over a metal sample with an inverted test-tube to collect any gas produced then some discussion may be required.

Background theory

Knowledge of the reactions of Group 1 metals with water for comparison.

Safety

Wear eye protection.

The transition metal compounds may be harmful or irritant, as may their solutions, depending on the concentration.

Ammonia vapour irritates eyes, lungs and the respiratory system

Answers

1. They are hard, dense and shiny. They are good conductors of heat and electricity. They are also malleable and ductile.
2. Transition metals react with water very slowly, if at all.
3. As well as the above they also form coloured compounds. They form compounds that can have more than one formula.

Properties of the transition metals and their compounds

Introduction

The Periodic Table allows chemists to see similarities and trends in the properties of chemical elements. This experiment illustrates some properties of the common transition elements and their compounds.

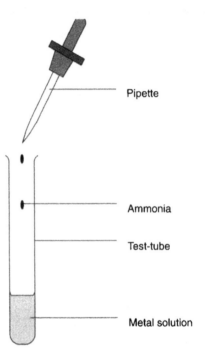

What to record

What was observed. A table may be useful.

What to do

1. Test the metal samples for hardness and ability to bend without breaking. Complicated apparatus is not needed for this! Record your answers qualitatively.
2. Find out which samples are magnetic.
3. Set up an experiment to see if the metals react with water. (This may need to be left for some time).
4. Take a small sample of a solution of copper(II) sulfate (approximately 2 cm^3), add ammonia solution to it a few drops at a time. Record your observations. Add ammonia solution until there is no further change.
5. Repeat with the other solutions of transition metal compounds.

Safety

Wear eye protection.

The transition compounds may be harmful or irritant.

Ammonia vapour irritates eyes, lungs and the respiratory system.

Questions

1. Describe the physical properties of transition metals.
2. How do transition metals react with water?
3. What properties do the compounds of transition metals have in common?

RS•C

89. Halogen compounds

Topic

The Periodic Table, reactivity within a Group, solubility.

Timing

60 min (depending on how much is attempted).

Description

This experiment is designed to indicate some properties of the hydrogen halides and halide ions. In this experiment students discover the acid properties of hydrogen halides, the order of reactivity, the qualitative tests for halide ions, and the differing solubilities of their lead salts. A description of the fountain experiment can be found in *Classic Chemistry Demonstrations*, p. 214. London: RSC, 1995.

Apparatus and equipment (per group)

- ▼ Three test-tubes
- ▼ Three boiling tubes
- ▼ 250 cm^3 Beaker
- ▼ Glass rod.
- ▼ White tile.

Chemicals (per group)

- ▼ Dilute hydrochloric acid — 0.4 mol dm^{-3}
- ▼ Dilute sodium hydroxide — 0.4 mol dm^{-3} (**Irritant**)
- ▼ Aqueous solutions of the chloride, bromide and iodide of sodium or potassium (approximately 0.2 mol dm^{-3}). (Suitable solutions are: potassium chloride 1.5 g in 100 cm^3, potassium bromide 2.4 g in 100 cm^3, and potassium iodide 3.5 g in 100 cm^3)
- ▼ Lead nitrate solution (approximately 0.2 mol dm^{-3}, 6.5g in 100 cm^3) (**Toxic**)
- ▼ Aqueous chlorine (**Toxic vapour**)
- ▼ Silver nitrate — 0.1 mol dm^{-3}.

Teaching tips

Students should rinse dirty test-tubes with deionised water before using silver nitrate.

Background theory

Students should be aware of the ionic nature of the compounds being investigated.

Safety

Wear eye protection.

Halogen compounds

Introduction

The halogens are elements of Group 7 of the Periodic table. This experiment illustrates some of the trends and similarities within the compounds of this group.

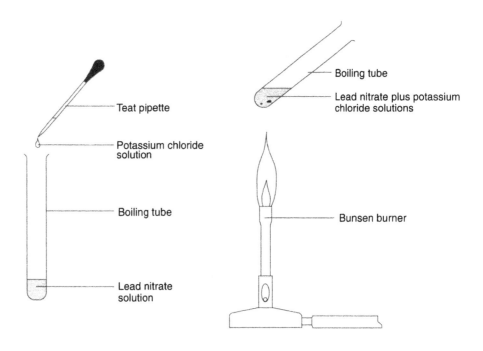

Safety

Wear eye protection.

What to record

What was observed. The table may be used.

What to do

Test 1 Silver nitrate	Observations
Put a little sodium chloride solution (Cl^- ions) in a test-tube and add five drops of silver nitrate solution. Leave in the light for a few minutes.	
Put a little sodium bromide solution (Br^- ions) in a test-tube and add five drops of silver nitrate solution. Leave in the light for a few minutes.	
Put a little sodium iodide solution (I^- ions) in a test-tube and add five drops of silver nitrate solution. Leave in the light for a few minutes.	

Test 2 Chlorine water (Toxic vapour)	Observations
Add a few drops of a solution of chlorine (**Toxic vapour**) to sodium bromide solution.	
Add a few drops of a solution of chlorine (**Toxic vapour**) to sodium iodide solution.	

Test 3 Lead nitrate (Toxic)		Observations
Put approximately 4 cm^3 of lead nitrate solution (**Toxic**) into a boiling tube, then add five drops of potassium chloride solution to a boiling tube and heat till it boils.	Place all three tubes in a beaker of cold water to cool.	
Put approximately 4 cm^3 of lead nitrate solution (**Toxic**) into a boiling tube, then add five drops of potassium bromide solution to the boiling tube and heat till it boils.		
Put about 4 cm^3 of lead nitrate solution (**Toxic**) into a boiling tube, then add five drops of potassium iodide solution to a boiling tube and heat till it boils.		

90. Finding the formula of an oxide of copper

Topic

Moles, stoichiometry and formulae.

Timing

60 min + further time if a spreadsheet is used to analyse class results.

Description

Students reduce copper(II) oxide to copper. The mass is recorded before and after the reaction. The ratio of copper to oxygen in copper(II) oxide can be calculated.

Apparatus and equipment (per group)

- ▼ Hard glass test-tube with small hole near closed end
- ▼ Bung and glass tube to fit the test-tube, with rubber tubing attached to the glass tube for connecting to the gas tap
- ▼ Access to a top pan balance accurate to at least 0.01 g
- ▼ Retort stand
- ▼ Boss
- ▼ Clamp.

Chemicals (per group)

For best results use analytical grade copper(II) oxide which has been dried by heating in an open dish at 300–400 °C for 10 min and then stored in a dessicator.

Teaching tips

The copper oxide must be dry and very strong heating is required. It is essential to keep a small stream of gas passing through the apparatus until it is cold, otherwise the copper will re-oxidise.

Students should be told to gently shake the copper oxide as it is being reduced to expose the lower layers to the gas.

See also experiment No 67. Demonstration of the procedure would be helpful.

Background theory

Students need to be aware of the definition of a mole and how to calculate the number of moles from reacting masses.

Safety

Wear eye protection. Copper(II) oxide is harmful.

Extension

The class results can be processed using a spreadsheet. Similar calculations on other substances can be carried out from given data.

Finding the formula of an oxide of copper

Introduction

The chemical formula gives the types of atom in the substance. It also gives the relative number of each type. From the mass of each element in a sample, the number of moles can be calculated. The lowest whole number ratio provides the simplest chemical formula.

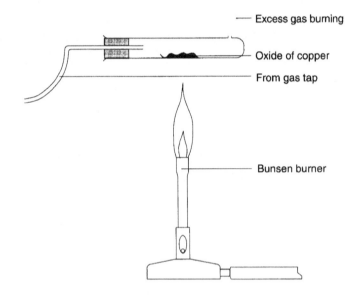

What to record

- Weight of test-tube + bung.
- Weight of test-tube + bung + copper(II) oxide.
- Weight of test-tube + bung + copper.

What to do

1. Weigh the test-tube + bung.
2. Place two spatulas of dry black copper(II) oxide in the centre of the tube. Try to spread it out.
3. Weigh the tube + bung + copper(II) oxide.
4. Assemble the apparatus as shown in the diagram.
5. Pass a gentle stream of gas through the tube without lighting it. This will flush out the air. After a few seconds set light to the gas and adjust the height of the flame coming out of the test tube to about 3 cm. Keep your head well back.
6. Heat the copper(II) oxide strongly and move the flame slowly to and fro. Continue to heat for five min after the solid has turned a brownish pink colour.
7. Stop heating the tube but keep the gas flowing through the test-tube and burning at the end. This prevents re-oxidation of the copper.
8. Let the test-tube cool, turn off the gas and reweigh the tube + bung + copper.

Safety

Wear eye protection

Copper(II) oxide is harmful.

Questions

1. What is the mass of copper(II) oxide used?
2. What is the mass of copper formed?
3. What is the mass of oxygen lost?
4. How many moles of copper were formed?
5. How many moles of oxygen were combined with this number of moles of copper?
6. What is the simplest whole number ratio of moles of copper to moles of oxygen?
7. What is the formula of copper(II) oxide?

RS•C

91. Making a fertiliser

Topic

Ammonia, the Haber process, industrial chemistry.

Timing

60 min.

Description

This experiment involves preparing ammonium sulfate. This is an effective fertiliser.

Apparatus and equipment (per group)

- Evaporating basin
- Gauze
- Tripod
- Bunsen burner
- 20 cm^3 Measuring cylinder
- Filter paper
- Filter funnel
- Conical flask (to stand funnel on)
- Glass rod.

Chemicals (per group)

- Sulfuric acid   1 mol dm^{-3} (**Corrosive**)
- Ammonia solution   2 mol dm^{-3}
- Full range indicator paper.

Teaching tips

Students may need to be told that a way of checking the pH of their solutions is to take a drop on a glass rod and place it on a piece of full range indicator paper.

Background theory

Students should understand that ammonia is an alkali and neutralises acid. They should also be aware that crystals can be obtained by evaporating a solution and leaving it to cool.

Safety

Wear eye protection.

The ammonia solution gives off ammonia which irritates eyes, lungs and respiratory system.

Sulfuric acid causes burns.

Answers

1. Ammonia + sulfuric acid → ammonium sulfate + water
2. $2NH_4OH + H_2SO_4 \rightarrow (NH_4)_2SO_4 + 2H_2O$
3. 21 per cent nitrogen.

Making a fertiliser

Introduction

Producting fertilisers is very important. This experiment involves preparing ammonium sulfate. Ammonium sulfate is a popular and effective fertiliser.

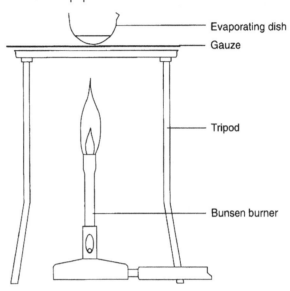

What to record

What was done and what was observed.

What to do

1. Put 20 cm^3 sulfuric acid into an evaporating basin.
2. Add the ammonia solution a little at a time, with stirring, until a definite smell of ammonia is obtained.
3. Check the pH is 7 or above with indicator paper.
4. Evaporate the solution to about one-fifth of its original volume (**Care** – do not let the solution spit), and cool.
5. Filter off the crystals and dry.

Safety

Wear eye protection.

The ammonia solution gives off ammonia which irritates eyes, lungs and respiratory system.

Sulfuric acid causes burns.

Questions

1. Write a word equation for this preparation.
2. Write a balanced symbol equation for this preparation.
3. Calculate the percentage of nitrogen in this fertiliser.

RS•C

92. Electrolysing copper(II) sulfate solution

Topic

Electrolysis, metal extraction, electroplating.

Timing

45 min.

Description

Students electrolyse copper(II) sulfate solution using graphite electrodes. Students identify the cathode as the electrode with the copper deposit on it then exchange this carbon cathode for an item of their choice (key ring, door key *etc*). Students can then try to remove the copper plate by reversing the polarity.

Apparatus and equipment (per group)

- Low voltage DC supply (3–6 V)
- Two leads
- Two crocodile clips
- 100 cm^3 Beaker.

Chemicals (per group)

- Two graphite electrodes
- Copper(II) sulfate solution　　　　　　　　　　　　　　　　0.1 mol dm^{-3} (**Harmful**).

Teaching tips

Students should understand the terms anode, cathode, electrode and electrolyte, so this may need to be explained. If students are unsure of electrical circuit diagrams, they may need some explanation.

Safety

Wear eye protection.

Answers

1. Cathode (negative electrode).
2. The copper must have a positive charge in the solution.
3. Using dirty copper as the anode and clean copper as the cathode.

Electrolysing copper(II) sulfate solution

Introduction

When a solution contains positive and negative ions, it conducts electricity. When electricity passes through a solution, chemical reactions may occur at the electrodes. This process is called electrolysis.

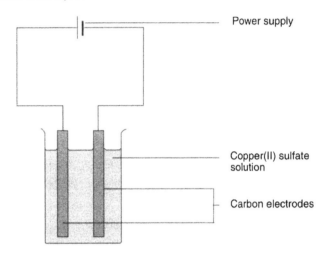

What to record

Which electrode has a fresh deposit of copper?

What to do

1. Set up the circuit shown in the diagram. Do not switch on the power supply until everything is ready.
2. Switch on and leave for about 5 min. Watch what happens and record your results.
3. Exchange the copper plated electrode for a key or key ring, or other article of your choice. Leave for 5 min.
4. Attempt to remove the copper plate by reversing the polarity.

Safety

Wear eye protection.

Copper(II) sulfate is harmful.

Questions

1. To which electrode has the copper gone?
2. What does this indicate about the charge on the copper in solution?
3. Suggest a way to use this process to purify impure or scrap copper.

RS•C

93. Producing a foam

Topic

Suspension (gas in liquid).

Timing

20 min.

Description

Two solutions are mixed and a chemical foam is produced.

Apparatus and equipment (per group)

- Two 250 cm^3 beakers
- Pestle
- Mortar.

Chemicals (per group)

- Laundry detergent – eg Persil non-biological (**Care** – may be irritant)
- Hydrated aluminium sulfate
- Sodium hydrogen carbonate.

Teaching tips

It is sometimes difficult to dissolve these reagents, but the reaction still works. The foam is produced by the action of carbon dioxide gas on a detergent solution.

Hydrogen carbonate ions from the detergent react with hydroxonium ions from the aluminium sulfate and water to produce carbon dioxide.

$HCO_3^-(aq) + H_3O^+(aq) \rightarrow 2H_2O(l) + CO_2(g)$

Teachers need to tell students that aluminium sulfate in water produces H_3O^+ ions.

This is similar to the baking process; baking powder contains sodium hydrogen carbonate and tartaric acid.

Background theory

This chemical foam contains carbon dioxide (CO_2), while mechanical foams often contain air.

This foam is a colloidal system with a gas dispersed in a liquid. This is a suspension of gas in the liquid.

Other common foams include whipped cream and shaving cream.

Safety

Wear eye protection

Answers

1. $HCO_3^-(aq) + H_3O^+(aq) \rightarrow 2H_2O(l) + CO_2(g)$
2. Baking powder contains sodium hydrogen carbonate and tartaric acid.
3. Whipped cream and shaving cream.

Producing a foam

Introduction

This experiment examines how a stable foam is produced and what reactions are involved.

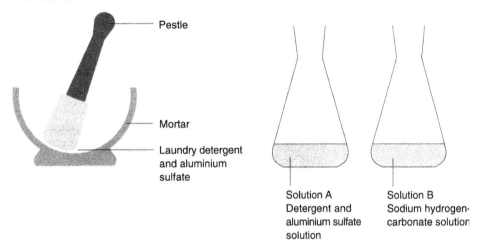

What to record

What happened (observations).

What to do

1. Put 1 g of laundry detergent and 7 g of aluminium sulfate in a mortar and grind into a fine powder with a pestle.
2. Dissolve this powder in approximately 50 cm^3 of water in a conical flask (A).
3. Dissolve 5 g of sodium hydrogen carbonate in 50 cm^3 water in another conical flask (B).
4. Pour the contents of flask A into flask B and mix quickly.

Safety

Wear eye protection.

Detergent may irritate the skin.

Questions

1. What reactions lead to the production of the foam?
2. How is this reaction similar to that involving the production of carbon dioxide (CO_2) during the baking process?
3. Name some other examples of foams.

RS•C

94. Getting metals from rocks

Topic

Extraction of metals, reactivity series.

Timing

40 min.

Description

Students heat a mixture of copper carbonate (**Harmful**) and charcoal in a crucible. The hot mixture is poured into water. The products are washed with water and copper metal is obtained.

Apparatus and equipment (per group)

- ▼ Crucible and pipe clay triangle (or the lid of a tin)
- ▼ 250 cm^3 Beaker
- ▼ Tongs.

Chemicals (per group)

- ▼ Malachite ore – powdered or in very small pieces, or copper(II) carbonate powder
- ▼ Charcoal powder
- ▼ Charcoal in small pieces (crushed barbecue charcoal).

Teaching tips

The ore (if used) needs to be well crushed.

Strong heating is essential. Teachers should decide whether to allow students to heat the pile from above as well as below. The layer of carbon granules reduces the loss of powder during heating as well as reducing the amount of air reaching the centre of the pile. Pouring the hot mixture into water can reduce the re-oxidation of the copper formed, but this requires care.

Background theory

Students should be aware of the idea of the competition of elements.

Safety

Wear eye protection. Be aware of the risks from sparks (especially if barbecue charcoal is used).

Answers

1. Copper carbonate → copper oxide + carbon dioxide
   $CuCO_3 \rightarrow CuO + CO_2$
2. Copper oxide + carbon → copper + carbon dioxide
   $2CuO + 2C \rightarrow 2Cu + CO_2$
3. $CuCO_3 + H_2SO_4(aq) \rightarrow CuSO_4(aq) + CO_2 + H_2O$
   At cathode:
   $Cu^{2+} + 2e^- \rightarrow Cu$

Getting metals from rocks

Introduction

This experiment involves producing copper from copper ore (malachite). The composition of malachite is mainly copper carbonate ($CuCO_3$). This experiment involves heating the copper carbonate with carbon.

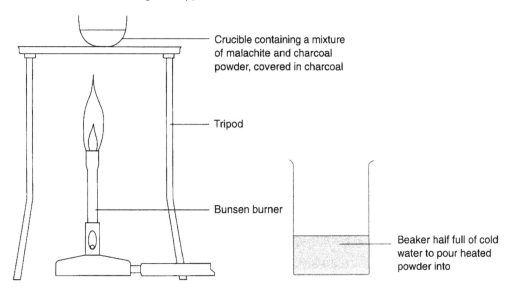

What to do

1. Mix a spatula of crushed malachite with two spatulas of charcoal powder. (This can be done in a dry test-tube or on a piece of paper).
2. Make a pile of the mixture in a crucible and then cover the pile with a layer of charcoal in small pieces.
3. Put the crucible on a tripod and pipe clay triangle and heat very strongly. (**Beware sparks**)
4. Half fill the beaker with water and then use tongs to tip the powder from the crucible into the water.
5. Swirl the beaker round so any copper falls to the bottom and then pour off the water and charcoal.
6. Add more water and keep on pouring and swirling so only the heavy material is left at the bottom of the beaker.

Safety

Wear eye protection.

Copper compounds are harmful.

Beware sparks.

Questions

1. Malachite ore is mainly copper carbonate ($CuCO_3$). When heated it produces carbon dioxide.
   Write word and symbol equations for the decomposition of copper carbonate.
2. Carbon reduces copper oxide to copper and carbon dioxide.
   Write word and symbol equations for this reaction.
3. An alternative method of extracting copper from malachite is by reaction with sulfuric acid, followed by electrolysis.
   Write formula equations for this reaction scheme.

95. Addition polymerisation

Topic

Organic chemistry, alkenes and polymers.

Timing

60 min.

Description

Phenylethene (styrene) is polymerised using a free-radical initiator.

Apparatus and equipment (per group)

▼ Boiling tube fitted with bung and 20 cm length of glass tubing
▼ Beaker of boiling water as water bath
▼ 100 cm^3 Beaker.

Chemicals (per group)

▼ Phenylethene (**Harmful, flammable**)
▼ Di(dodecanoyl)peroxide (lauroyl peroxide) (**Irritant, oxidising**)
▼ Ethanol 50 cm^3 (industrial methylated spirits) (**Highly flammable**)

Most styrene samples contain an inhibitor, which needs to be removed by washing with 1 mol dm^{-3} sodium hydroxide solution (**Corrosive**), then water, in a separating funnel. This then needs drying with anhydrous sodium sulfate for 10 min.

Teaching tips

This experiment is only suitable for extremely able students.

Teachers should be aware that addition polymerisation is difficult to demonstrate, or for students to experience themselves.

The amount of initiator used is very small.

The problem with this experiment is the comparison of the product with the commercial product. Students could consider why this is so.

Background theory

Students should be aware of the addition reactions of alkenes.

Safety

Wear eye protection.

Ideally use a fume cupboard at least for dispensing liquid. If not, good ventilation is essential as styrene vapour is narcotic in high concentrations.

Addition polymerisation

Introduction

Alkenes (carbon compounds containing double bonds) undergo addition reactions. In this experiment molecules of phenylethene (styrene) – the monomer – add on to each other to form polyphenylethene (polystyrene) - the polymer.

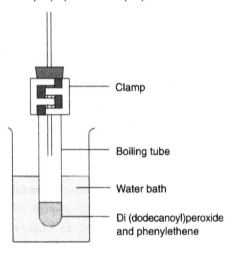

What to record

What was done. Compare the appearance of your product with the starting material.

What to do

1. Prepare a beaker of boiling water to act as a water bath. Keep your Bunsen burner away from all other chemicals.
2. Add 0.1 g of di(dodecanoyl) peroxide (**Irritant**) to 5 cm^3 of phenylethene (**Flammable**) in a boiling tube.
3. Put a bung containing a 20 cm length of glass tubing in the top and clamp the tube in the boiling water bath.
4. Heat for about 30 min and leave to cool. Extinguish all flames.
5. Pour the contents of the tube into 50 cm^3 of ethanol (**Highly flammable**). Use a glass rod to push the polyphenylethene into a lump and pour off the ethanol.
6. Dry the solid on a filter paper.

Safety

Wear eye protection. Work in a fume cupboard or ensure good ventilation.

96. Cracking hydrocarbons

Topic

Oil, petroleum industry, polymerisation, chemistry of alcohols.

Timing

60 min.

Description

The molecules to be cracked are vaporised and passed over a heated catalyst. The gas produced is collected over water. This experiment is a good test of manipulative skill and observation.

Apparatus and equipment (per group)

- ▼ Hard glass test-tube or boiling tube
- ▼ Delivery tube with bung to fit this tube
- ▼ Bunsen valve
- ▼ Trough or bowl
- ▼ Three test-tubes with bungs to collect gas.

Chemicals (per group)

- ▼ One of: liquid paraffin, decane, polythene (polyethene), ethanol (**Highly flammable**)
- ▼ Mineral wool
- ▼ Aluminium oxide granules (or broken porcelain chips)
- ▼ Bromine water 0.04 mol dm^{-3} (**Harmful, irritant**)

Teaching tips

Use of a Bunsen valve:

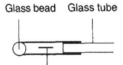

Rubber tube with slit in it

Gas can pass through the slit from the inside but if the pressure drops inside the sides of the slit are pushed together.

Even with a Bunsen valve, suck back may happen. Students should be told to lift the stand to raise the delivery tube out of the water at the end of the experiment. Teachers should check the students' apparatus for blockages caused by melted bungs or catalyst blocking the delivery tube out of the test-tube, since explosions can occur.

Background theory

Students need to know the arrangement of the molecules in the reactant and understand that the catalyst speeds up the reaction. Knowledge of the properties of alkenes is not necessary before the reaction.

Safety

Wear eye protection.

1,2-Dibromoethane is harmful, but very little forms.

RS•C

Cracking hydrocarbons

Introduction

The demand for petrol is greater than the amount produced by distilling crude oil. The cracking of hydrocarbons also produces molecules which can be converted into petrol. This experiment models the industrial cracking process.

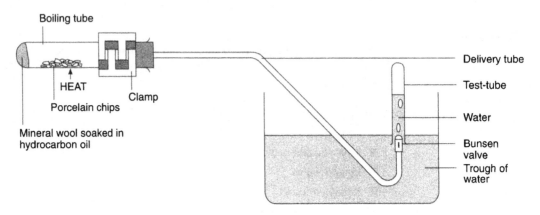

What to record

Record:

1. The appearance of the original oil and the gas.
2. Whether the oil and gas burn.
3. Whether the oil and gas smell.
4. The effect of the oil and the gas on aqueous bromine water.
5. A table may be helpful.

What to do

1. Set up the apparatus as shown in the diagram. Fill four test-tubes with water and invert them in the trough.
2. Strongly heat the catalyst (porcelain chips) for a few minutes.
3. Now flick the flame onto the end of the tube containing the mineral wool and the oil. Try to produce a steady stream of bubbles.
4. Collect tubes of the gas. Discard the first one, which will be mainly air.
5. Stopper the three tubes of gas and test them as follows
   (a) What do they look like?
   (b) What do they smell like (care)?
   (c) Use a lighted splint to see if they burn.
   (d) Add two drops of aqueous bromine and shake.

Safety

Wear eye protection.

Be careful to avoid the water in the trough sucking back. When heating stops lift the apparatus out of the water using the stand.

Make sure the apparatus is not blocked. If no gas appears check the bung has not melted and that the catalyst has not blocked the delivery tube.

Aqueous bromine water is harmful and irritant.

Classic chemistry experiments

RS•C

97. Displacement reactions between metals and their salts

Topic

Reactivity series, displacement reactions, salts, transition metals.

Timing

40 min.

Description

Students can predict the reactivity series from this experiment. Alternatively, the results can be predicted using the reactivity series. Using plastic spotting tiles minimises contamination and reduces the quantities of material required.

Apparatus and equipment (per group)

- ▼ Plastic spotting tile – these should be clean
- ▼ Emery paper
- ▼ Dropping pipette. Use the type of teat pipette (usually fitted to Universal Indicator bottles) that does not allow squirting – *eg* Griffin
- ▼ Labels or felt pen that will write on spotting tiles.

Chemicals (per group)

- ▼ Five pieces approximate size 10 mm x 5 mm of:
  - strips of zinc foil
  - strips of magnesium ribbon
  - strips of copper foil
  - strips of lead foil

- ▼ access to
  zinc nitrate solution   0.1 mol dm^{-3}
  magnesium nitrate solution,   0.1 mol dm^{-3}
  copper nitrate solution,   0.1 mol dm^{-3}
  lead nitrate solution,   0.1 mol dm^{-3} (**Toxic**).

Teaching tips

To prevent contamination it is a good idea for students to put the solutions onto the spotting tiles sequentially (students can label each row).

Students may need guidance as to whether a reaction has occurred. This requires careful observation of the metal and the solution.

See also *Microscale Chemistry*, p. 24. London: RSC, 1997.

RS•C

Background theory

Students may be aware of competition reactions for oxygen, which could be a starting point.

Safety

Wear eye protection. Lead nitrate is toxic.

Answers

1. Magnesium, zinc, lead and copper

Displacement reactions between metals and their salts

Introduction

Some metals are more reactive than others. In this experiment, a strip of metal is added to a solution of another. If the metal is more reactive than the metal in solution, this metal displaces (pushes out) the less reactive metal from the solution.

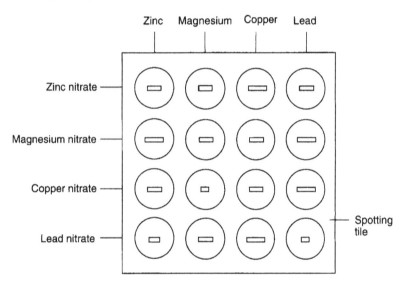

What to record

Record which metals react with the solutions.

A table may be useful. Use a ✔ to show reactivity and an ✘ to show no reaction.

Metal / Solution	Zinc nitrate solution	Magnesium nitrate solution	Copper nitrate solution	Lead nitrate solution
Zinc				
Magnesium				
Copper				
Lead				

What to do

1. Clean each of the metal strips with emery paper.
2. Using a teat pipette put some of the solution of a metal compound in four of the holes in the spotting tile. (Label this row with the name of the solution).
3. Do this for each solution of a metal compound.
4. Put a piece of each metal in each of the solutions.
5. Put a tick or a cross in your table to show if they have reacted.

Safety

Wear eye protection. Some solutions are toxic.

Questions

1. What is the order of reactivity of these metals (from most to least reactive)?

98. The effect of temperature on solubility

Topic

Materials, solubility.

Timing

60 min.

Description

Students heat water with a solute until it dissolves. The solution is then cooled until crystallisation occurs. More water is added. The solution is again heated until the crystals dissolve. The new temperature when crystals appear is recorded.

Apparatus and equipment (per group)

- Boiling tubes
- 250 cm^3 Beaker to act as ice bath.
- 250 cm^3 Beaker to act as a hot water bath.
- Stirring thermometer (-10–110 °C)
- 10 cm^3 Measuring cylinder or graduated pipette.
- Wooden tongs to hold hot boiling tube.

Chemicals (per group)

- Ammonium chloride (**Harmful**)
- Ice (crushed or small pieces).

Teaching tips

This is a good opportunity to introduce the use of quantitative chemical apparatus to younger students.

Background theory

Students should know that solids are generally more soluble in hot water than in cold water.

Safety

Wear eye protection.

The effect of temperature on solubility

Introduction

Most solid substances that are soluble in water are more soluble in hot water than in cold water. This experiment examines solubility at various temperatures.

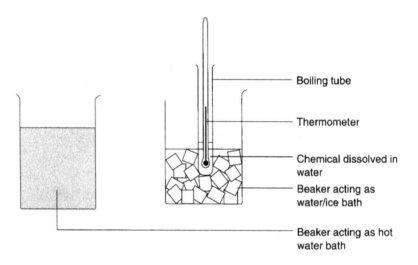

What to record

Fill in the temperatures.

Volume of water/cm^3	Solubility/g dm^3	Crystallisation temperature /° C
4	650	
5	520	
6	433	
7	371	
8	325	
9	289	
10	260	

(The crystallisation temperature is the temperature at which crystals appear).

What to do

1. Set up a hot water bath and an ice bath. Put 2.6 g of ammonium chloride (**Harmful**) into the boiling tube. Add 4 cm^3 water.
2. Warm the boiling tube in the hot water bath until the solid dissolves.
3. Put the boiling tube in the ice bath and stir with the thermometer. Use wooden tongs to hold it if necessary.
4. Note the temperature at which crystals first appear and record it in the table
5. Add 1 cm^3 water. Warm the solution again, stirring until all the crystals dissolve.

6. Then repeat the cooling and note the new temperature at which crystals appear.
7. Repeat steps 5, 6 and 7 until 10 cm^3 water has been used.

Safety

Wear eye protection.

Questions

1. Plot a graph showing solubility on the vertical axis and temperature on the horizontal axis.

RS•C

99. Purifying an impure solid

Topic

Mixtures, separation methods, purification.

Timing

60 min.

Description

A dirty looking solid is given to students. Students attempt to obtain the pure substance by dissolution, filtration, evaporation and crystallisation.

Apparatus and equipment (per group)

- ▼ 100 cm^3 Beaker
- ▼ Evaporating basin
- ▼ Filter funnel and filter paper
- ▼ Glass rod.

Chemicals (per group)

- ▼ Sample of crude alum (add dry soil to alum crystals).

Teaching tips

If students have not seen this technique, it may be worthwhile starting with a class discussion. Suggestions as to how to remove the dirt could then be quickly tried as a cooperative demonstration.

If teachers want their students to try to grow large crystals they should tell them to retain the liquid above the crystals (a saturated solution of alum), and suspend a crystal in it.

Ensure students use only the minimum amount of water or the experiment does not work.

Background theory

Students should understand that some substances are soluble in a particular solvent (in this case water) and others are not. They should also understand that crystals form because the solubility of alum is much less in cold water than it is in hot water.

Safety

Wear eye protection. Take care with hot apparatus.

Purifying an impure solid

Introduction

It is often necessary to obtain a pure chemical from an impure sample. This experiment involves the purification of a chemical called alum.

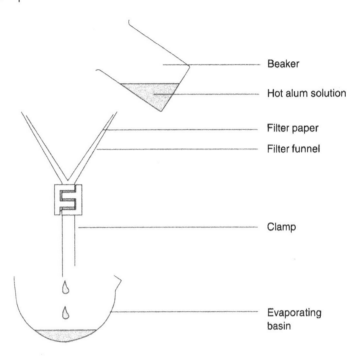

What to record

What was done.

What to do

1. Cover the bottom of the beaker with alum.
2. Add enough water to cover it.
3. Heat and stir the mixture with a glass rod.
4. Wait until the mixture boils. If the alum has not dissolved, (ignore the dirt) add some more water and boil.
5. Filter the hot solution into the evaporating basin.
6. Leave the solution to cool.
7. Pour off the liquid and dry the crystals on filter paper.

Safety

Wear eye protection. Take care with hot apparatus.

Questions

1. Write a couple of sentences to describe this experiment. Make sure all of the following words are used:
dissolve, solvent, filter, filtrate, evaporate, solute, crystallise, crystal.

RS•C

100. Chemicals from seawater

Topic

Mixtures and separation.

Timing

60 min.

Description

Students reduce the volume of seawater by boiling. Different salts are obtained. This illustrates that seawater is not a single substance.

Apparatus and equipment (per group)

- ▼ 250 cm^3 Beaker
- ▼ 100 cm^3 Beaker
- ▼ Filter funnel and filter paper
- ▼ Tripod
- ▼ Gauze
- ▼ Bunsen burner.

Chemicals (per group)

- ▼ 200 cm^3 seawater

To produce seawater: bubble CO_2 through a mixture of 250 cm^3 limewater/ 750 cm^3 deionised water for about 20 min or until the cloudy precipitate disappears completely. Filter. Add as much hydrated calcium sulfate as will dissolve. Add sodium chloride. Decant the liquid.

- ▼ Access to dilute hydrochloric acid   1 mol dm^{-3}.

Teaching tips

Remind students to record what they observe. One teacher suggested the suitability of this experiment early in a science course while using scientific apparatus is still a novelty. It provides practice at using apparatus but also introduces mixtures and some simple chemical tests.

Teachers may wish to mark beakers at the 60 cm^3 mark.

Background theory

Students should understand that some substances are soluble in water.

Safety

Take care with hot glassware and solutions. Wear eye protection.

Answers

1. No.
2. The various salts have different solubilities in water.

Chemicals from seawater

Introduction

Seawater is often called salt water. Seawater contains various different salts. This experiment involves separating some of these salts from the mixture.

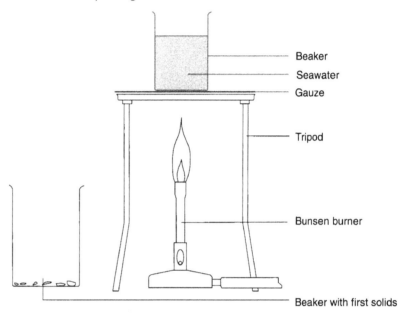

What to record

What was done and what was seen.

What to do

1. Put 200 cm^3 of seawater into a 250 cm^3 beaker.
2. Heat the seawater to boiling point and boil the liquid until 60–70 cm^3 remains. Stop heating when solid is observed and let the solid settle.
3. Pour (decant) the clear liquid into a 100 cm^3 beaker.
4. Add a few drops of hydrochloric acid to the solid left behind. What happens?
5. Put the beaker back on the heat and boil it again until another solid appears (probably when the liquid level is between 20 cm^3 and 40 cm^3.)
6. Filter the liquid off, wash out the beaker, and boil the liquid again until there is almost none left. Let it cool. What is observed?

Safety

Wear eye protection. Take care with hot apparatus and solutions.

Questions

1. Is seawater a single substance?
2. What is the difference between these separated salts?